Origin of Thermoremanent Magnetization

ADVANCES IN EARTH AND PLANETARY SCIENCES

General Editor:

T. RIKITAKE (Tokyo Institute of Technology)

Editorial Board:

S. AKASOFU (University of Alaska)
S. AKIMOTO (University of Tokyo)
Y. HAGIWARA (University of Tokyo)
H. KANAMORI (California Institute of Technology)
C. KISSLINGER (University of Colorado)
A. MASUDA (University of Kobe)
A. NISHIDA (University of Tokyo)
M. OZIMA (University of Tokyo)
R. SATO (University of Tokyo)
S. UYEDA (University of Tokyo)
I. YOKOYAMA (Hokkaido University)

Special Issue of Journal of Geomagnetism and Geoelectricity

Origin of Thermoremanent Magnetization

Proceedings of AGU 1976 Fall Annual Meeting
December 1976, San Francisco

Edited by
David J. Dunlop

Center for Academic Publications Japan
Japan Scientific Societies Press

Published by:
CENTER FOR ACADEMIC PUBLICATIONS JAPAN
JAPAN SCIENTIFIC SOCIETIES PRESS
20-6 Mukogaoka, 1-chome, Bunkyo-ku, Tokyo 113, Japan

Sole distributor for the outside Japan:
BUSINESS CENTER FOR ACADEMIC SOCIETIES JAPAN
20-6 Mukogaoka, 1-chome, Bunkyo-ku, Tokyo 113, Japan

JSSP No. 01362-1104

Printed in Japan

Preface

Physical and chemical studies of the earth and planets along with their surroundings are now developing very rapidly. As these studies are of essentially international character, many international conferences, symposia, seminars and workshops are held every year. To publish proceedings of these meetings is of course important for tracing development of various disciplines of earth and planetary sciences though publishing is fast getting to be an expensive business.

It is my pleasure to learn that the Center for Academic Publications Japan and the Japan Scientific Societies Press have agreed to undertake the publication of a series "Advances in Earth and Planetary Sciences" which should certainly become an important medium for conveying achievements of various meetings to the academic as well as non-academic scientific communities. It is planned to publish the series mostly on the basis of proceedings that appear in the Journal of Geomagnetism and Geoelectricity edited by the Society of Terrestrial Magnetism and Electricity of Japan, the Journal of Physics of the Earth by the Seismological Society of Japan and the Volcanological Society of Japan, and the Geochemical Journal by the Geochemical Society of Japan, although occasional volumes of the series will include independent proceedings.

Selection of meetings, of which the proceedings will be included in the series, will be made by the Editorial Committee for which I have the honour to work as the General Editor. I and the members of the Editorial Committee will certainly welcome any suggestions that will promote the series. Whenever the convener of a meeting related to earth and planetary sciences is in a position to have to look for a medium for publishing the proceedings please contact us.

Tsuneji Rikitake
General Editor

Foreword

Thermoremanent magnetization (TRM) is of central importance to paleomagnetism, but its mechanism and even its diagnosis in nature have remained uncertain. On December 9, 1976, a full-day session on 'The Origin of TRM' was held as part of the fall annual meeting of the American Geophysical Union in San Francisco. Of the six invited and sixteen contributed papers presented at the session (for abstracts, see *EOS*, **57**, 904–907, 1976), thirteen appear in final form in this special issue of the *Journal of Geomagnetism and Geoelectricity*.

The papers fall into two broad groups, the first incorporating theoretical and laboratory studies of TRM (traditional 'rock magnetism'), the second dealing with TRM in natural materials (traditional 'paleomagnetism'). As the papers themselves make clear, the artificial distinction between rock magnetism and paleomagnetism is gradually disappearing: the problems that remain to be solved, the origin of TRM among others, require a flexible point of view and attack on many fronts.

Six of the papers in this issue consider the physical origin of TRM in single-domain, pseudo-single-domain, and multidomain grains of magnetite and titanomagnetite. Three deal with other minerals: titanomaghemite, hematite, and iron. The identification of TRM and its role compared to other remanence mechanisms are considered for continental rocks in one paper, for submarine rocks and ophiolites in two papers, and for extraterrestrial materials and their analogs in two papers.

I am indebted to Dr. Richard Blakely for his help in organizing the session, to Dr. Minoru Ozima for invaluable editorial assistance, to Dr. John Verhoogen for contributing an introduction to this issue and to those who so ably and speedily reviewed or substantially commented on the papers: M.E. Bailey, S.K. Banerjee, A. Brecher, R.F. Butler, C.M. Carmichael, D.W. Collinson, R. Day, M.E. Evans, M. Fuller, W.A. Gose, R.B. Hargraves, H.P. Johnson, M. Lanoix, S. Levi, R.T. Merrill, G.W. Pearce, P.H. Reynolds, E.J. Schwarz, P.N. Shive, and F.D. Stacey.

David J. Dunlop
Guest Editor

Introduction

The recognition that many rocks carry a remanent magnetization that reliably reflects the direction, and in fewer rocks also the intensity, of the magnetic field prevailing at the time and place of their formation, has lead in the past 25 years to spectacular developments in several branches of geology and geophysics. The discovery of temporal variations of the earth's magnetic field on a time scale of 10^3–10^7 years, and in particular of frequent reversals of its polarity, has lead to new ideas regarding the origin of the field and the behavior of the earth's core. Paleomagnetism finally provided, after years of debate, the clinching arguments for continental drift, and lead to the revolutionary concepts of sea-floor spreading and plate tectonics. Yet, paradoxically, the very mechanism by which rocks acquire their tell-tale remanence has remained somewhat obscure, particularly so in the case of the thermal remanent magnetization (TRM) acquired by igneous rocks as they cool in the earth's weak field. The two features of TRM that have made it so useful in paleomagnetism are its intensity and an extraordinary stability that enables it to survive over billions of years. The intensity of TRM in rocks is usually greater by several orders of magnitude than the remanence that can be imparted by exposure, at room temperature, to the earth's field; and destruction of TRM by a-c demagnetization may require an a-c field of several hundred gauss or more.

The theory of the acquisition of TRM has been satisfactorily developed, and experimentally tested, in the two extreme cases of 1) magnetic grains small enough to consist of single domains (SD), and 2) grains large enough to have a multi-domain (MD) structure. (The threshold for single-domain behavior in magnetite appears to be close to 0.05 μm for equidimensional grains.) It turns out that a field as weak as the earth's field may induce in an assembly of non-interacting single-domain grains a TRM comparable in magnitude to the saturation remanence induced at room temperature by fields of several thousand gauss, and of high stability. By contrast, weak-field TRM in multidomain grains is usually much weaker and less stable. A curious feature is that grains of size between 0.05 and about 15 μm (for magnetite), and which are clearly much too large to be SD, nevertheless exhibit an intensity of remanence and a resistance to demagnetization that are typical of SD grains. Such grains are said to show pseudo single-domain behavior (PSD). The TRM of PSD grains also shows an inverse dependence on grain size that is not typical of MD remanence, but the dependence of TRM on the magnitude of the inducing field is not that predicted for SD behavior.

PSD behavior is now generally attributed to the presence of residual moments that cannot be removed by demagnetization. Attempts to explain these residual moments by invoking internal strains (dislocations), inhomogeneity, or Barkhausen discreteness in domain-wall position, have generally failed to account for one or more characteristics of PSD, e.g., the inverse dependence of remanence on grain size. In the past 3 or 4 years, the search has focussed on the internal structure of domain walls ('walls within walls') and their termination at the surface of grains. The hunt, lead mainly by Stacey, Banerjee, and Dunlop, is still on, and the subject is lively enough to warrant publication, as a special issue of papers presented at a session on the origin of TRM held in December 1976 in San Francisco.

As the reader will judge, excellent progress is being made, but, as usual, much remains to be done before we can claim to fully understand the TRM of rocks. It is an unfortunate fact of nature that much of the TRM in rocks seems to be carried by grains so small (less than 1 μm) that they can hardly be seen under the microscope; neither their abundance, nor their size, shape or composition can be easily assessed. Nor can we, in many instances, determine precisely when they formed, as they may be products of chemical reactions (e.g., oxidation) occurring late in the cooling history of the rock. It is amusing to reflect that if the pioneering paleomagnetists of the early fifties had known, or even suspected, the full complexity (chemical, mineralogical, textural, magnetic-structural) of the magnetic properties of rocks, they probably would have thrown up their hands, declared rocks inherently unreliable, and turned to lesser things. Ignorance, it would seem, can sometimes be a blessing.

J. Verhoogen

CONTENTS

Adv. Earth Planet. Sci., **1**, 1–33, 1977

TRM and Its Variation with Grain Size*

Ron DAY

Department of Geological Sciences, University of California,
Santa Barbara, California, U.S.A.

(Received June 22, 1977)

Thermoremanent magnetization (TRM), the dominant mechanism in igneous rocks, has been investigated for many years, yielding a large data base of experimental results and several theoretical models. However, there are still a large number of discrepancies between the observations and the theories.

The theoretical models of TRM are reviewed, and then evaluated in the light of recent experimental results from sized synthetic magnetites and titanomagnetites, and igneous rocks.

1. Introduction

The fundamental assumption of paleomagnetism is that the direction and intensity of the natural remanent magnetization (NRM) of a rock are characteristic of the ambient geomagnetic field present during its formation. The combined facts that igneous rocks are magnetized by the geomagnetic field when they cool and that the TRM which they thus acquire is usually very stable, even over geological time, makes TRM a very important magnetic property of rocks. It has been known for centuries that rocks carried a remanent magnetization, but it was only during the last century (MELLONI, 1853) that it was established that the NRM contained a record of the ancient geomagnetic field. The basic characteristics of TRM emerged from the classical works of KOENIGSBERGER (1938), THELLIER (1938) and NAGATA (1941, 1942) on natural materials. Recognizing the difficulties and uncertainties involved with magnetic measurements on natural materials, a second generation of research began using synthetic powders (ROQUET, 1954; UYEDA, 1958; RIMBERT, 1959) and single crystals (SYONO, 1962). This was followed by experiments to determine the effects of stress (LOWRIE, 1967; SHIVE, 1969b) and magnetocrystalline anisotropy (OZIMA *et al.*, 1964; PETROV and METALLOVA, 1968). A third generation of work has been concerned with the effect of particle size on TRM (PARRY, 1965; ROBINS, 1972; DAY, 1973; DUNLOP, 1973a; RAHMAN *et al.*, 1973; LEVI, 1974; LEVI and MERRILL, 1976, 1977). The importance of this work is obvious; magnetic particles in rocks are known to span a large size range covering single-domain (SD), pseudo single-domain (PSD) and multi-domain (MD) behaviour, and it is well known that magnetic properties are very sensitive to changes in grain size (PARRY, 1965;

---

* Paper presented at the special session on the 'Origin of TRM,' American Geophysical Union, San Francisco, December 9, 1976.

DAY, 1973; DUNLOP, 1973a; LEVI, 1974; RAHMAN *et al.*, 1973; LEVI and MERRILL, 1976).

In spite of this wealth of data we have been unable to establish a satisfactory theoretical model of TRM. The models that have been proposed (NÉEL, 1949, 1955; STACEY, 1958; EVERITT, 1962; STACEY, 1962; DUNLOP, 1973a; SCHMIDT, 1973) are successful in explaining many of the general features of TRM, but some properties of TRM (i.e., TRM induction curves) still remain to be explained. In this paper, we will review the effects of grain size on the properties of TRM, and see how successful the proposed TRM models are in explaining the observed properties of TRM.

2. Transition Sizes in Titanomagnetites

The number of domains present in a magnetic particle depends primarily on the size and shape of the particle. Single-domain particles are generally characterized by large coercivities and intense stable remanences, because magnetic moment reversal occurs through coherent rotation, and are therefore natural candidates for carrying the TRM in rocks. However, below a certain transition size, d_{sp}, the particles become superparamagnetic, losing their remanence, while above a second transition size, d_o, the number of domains increases and again the intensity and stability of remanence will decrease. Thus, only those materials that have a wide single-domain range are likely to carry a stable SD remanence. Above the SD threshold the particle becomes subdivided by domain walls and the magnetic properties are then governed by domain wall motion.

Because it is relatively easy to move domain walls, multidomain (MD) particles are expected to have small coercivities and low remanence. There is evidence that this may not be true for small MD particles where only a few walls are present. STACEY (1962) has postulated the existence of pseudo single-domain (PSD) moments in small MD particles, with properties approaching those of SD particles. These PSD moments contribute to the TRM which we observe in rocks. It is therefore important to establish the size range for these three types of behaviour in the titanomagnetites.

2.1 Theoretical estimates of SD-MD transition size

The total energy of a magnetic particle is the sum of a number of physically distinct energy terms. In large particles the total energy is minimized by the formation of domains. However, in smaller grains it is possible that the lowest energy configuration is that of a single-domain grain. The method then is to compare the energy of a single-domain particle with the energy of the assumed domain configuration just above d_o. Several theoretical estimates of the SD transition size have been obtained. The methods differ only in the choice of domain configuration of the multi-domain particle. KITTEL (1949) calculated d_o in iron for a transition to a two domain particle and a four domain particle. He obtained sizes of the order of 10^{-6} cm. The same calculation for magnetite yields a transition size of about 6×10^{-6} cm.

These estimates turn out to be less than the domain wall thickness in both iron and magnetite. NÉEL (1947) proposed an alternative model, involving internal flux closure using a circular spin configuration. This gives a size of 3×10^{-6} cm for spherical iron. MORRISH and YU (1955) derived a similar model for the spinel structure and extended the calculations to include elongated grains (Table 1). Their derivation ignores magnetocrystalline energy which they indicate is reasonable for axial ratios less than 7 to 1. MURTHY *et al.* (1971) used an improved Kittel model for axial ratios greater than 7 to 1 (Table 1).

Micromagnetic theory (FREI *et al.*, 1957) provides a different approach in that it ignores the domain assumption completely and allows continuous spatial variation

Table 1. Transition sizes for titanomagnetites.

	Transition (sizes in microns)			
	SPM-SD	SD-PSD	a/b[1]	PSD-MD
A) Theory				
Magnetite				
MORRISH and YU (1955)		0.03	1	
		0.6	5	
MURTHY *et al.* (1971)		2–3		
EVANS (1972)	0.03–0.06	0.06[2]	1	
		0.4[2]	10	
BUTLER and BANERJEE (1975)	0.06	0.076	1	
		0.4	2.5	
		1.4	5	
Titanomagnetite				
DAY (1973)		0.1–0.3	1	
$X=0.6$[3]		0.7–4.0	5	
		1–20	10	
BUTLER and BANERJEE (1975)				
$X=0.3$	0.055	0.1	1	
		0.6	2.5	
$X=0.6$	0.08	0.3	1	
		1.3	2.5	
B) Experiment				
Magnetite				
DUNLOP (1973b)	0.03	0.05	~1	
PARRY (1965)				15–20
DAY (1973)		≦0.1		10–20
Titanomagnetite				
DAY (1973)				
$X=0.4$		~1[4]		
$X=0.8$		~1–2[4]		30–40
SOFFEL (1971)				
$X=0.55$		0.7		

[1] a/b is the axial ratio.
[2] Using FREI *et al.* (1957).
[3] Spread of values bracket estimated from several models.
[4] a/b is probably in range $1 < a/b < 3$.

of magnetization during reversal. However, the predicted thresholds for spherical and elongated magnetite do not differ significantly from the other approaches (EVANS, 1972).

DAY (1973) used the above models to estimate the threshold sizes in titanomagnetite. Table 1 shows that d_o increases with increasing titanium content. BUTLER and BANERJEE (1975) following the ideas of AMAR (1957, 1958a, b) refined the original Kittel model. In their two domain plus 180° wall configuration they included the magnetostatic energy of the free poles at the termination of the domain wall and the dependence of the domain wall energy on the wall width. When these effects are included a domain structure can develop in particles whose size is less than the equilibrium wall width in an extended medium. The transition sizes for titanomagnetites (Table 1) calculated using this model are probably the best estimates we have to date.

The superparamagnetic transition size, d_{sp}, can be calculated from NÉEL's (1955) relaxation equation (modified by BEAN and LIVINGSTON, 1959):

$$\tau = \frac{1}{f_0} \exp\left[E_B/kT\right]$$

where τ is the appropriate relaxation time, f_0 the frequency factor and E_B is the energy barrier opposing spontaneous reversal. For equant magnetite particles, E_B is derived from the magnetocrystalline anisotropy while for elongated particles the shape anisotropy will dominate. The calculated values of d_{sp} are given in Table 1.

2.2 *Experimental estimates of the SD-MD transition size*

The most obvious way to estimate the transition size is to employ the Bitter pattern technique which reveals the domain structure directly. Unfortunately the transition sizes in magnetite-rich titanomagnetites are well below the resolution of the optical microscope. However, SOFFEL (1971) has investigated the domain structure of the titanomagnetites ($X=0.55$) occurring in the Parkstein and Rauher Kulm

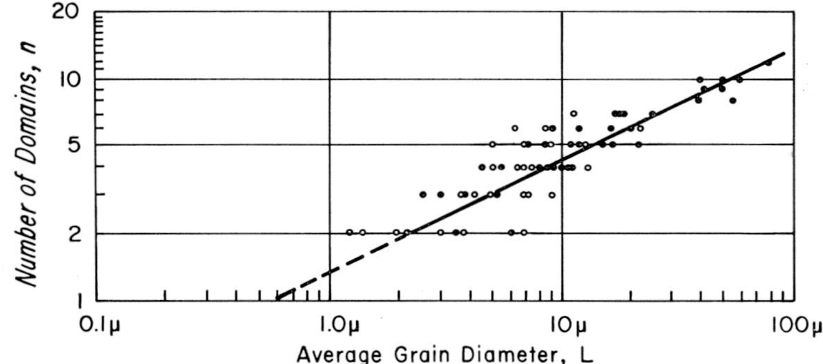

Fig. 1.   Plot of the average grain diameter L versus the number n of domains on a bilogarithmic scale for particles of various grain sizes in the Parkstein and Rauher Kulm basalt (from SOFFEL, 1971).

basalts. By plotting the number of domains versus the particle size (Fig. 1), he obtained a SD-MD transition size of 0.7 μm. Soffel's work shows two interesting features. First, curved domain walls are quite common [see, for example, Fig. 14 of SOFFEL (1971); also SOFFEL (1977), in this issue]. Second, the number of domains is not determined solely by the grain size. This is clearly illustrated in Fig. 1.

The transition size has been estimated indirectly from measurements of the magnetic properties of sized titanomagnetite particles. DUNLOP (1973b) estimated the transition sizes in magnetite from the hysteresis properties of four magnetite powders (mean particle sizes 0.037, 0.076, 0.10 and 0.22 μm). The SD-MD transition sizes in titanomagnetites ($X=0$ to $X=0.6$) were obtained from similar experiments by DAY (1973). These estimates are listed in Table 1.

The transition from PSD behaviour to true MD behaviour cannot be calculated. Further, it turns out that it is also difficult to determine the transition size experimentally because it depends on the particular magnetic property that is used for the estimate. After examining the variations of H_c,, H_{rc}, H_{rc}/H_c and J_r/J_s with increasing particle size. DAY (1973) estimated PSD-MD transition sizes of approximately 10–20 μm in magnetite ($X=0$) increasing to between 30–40 μm for $X=0.6$ titanomagnetite. A similar estimate for magnetite has been given by BAILEY (1975).

Table 1 summarizes our present knowledge of the titanomagnetite transition sizes.

3. Particle Size and Magnetic Properties

Saturation magnetization, Curie temperature and crystallographic cell size are not grain size dependent and for titanomagnetite they vary approximately linearly with composition. Consequently, they are used to identify the magnetic phases. Hysteresis properties do reflect differences in particle size, and can therefore be used to diagnose grain sizes. There are two effects to consider. First, there are the contrasting properties of SD, PSD and particles. Second, there are particle size effects within these boundaries.

DAY et al. (1976) compared the magnetic properties of SD and MD titanomagnetites. SD titanomagnetites possess large coercivities ($396 < H_c < 1910$ Oe, $560 < H_{rc} < 2140$ Oe), high remanence ($0.38 < J_r/J_s < 0.59$) and values of H_{rc}/H_c close to the

Table 2. Hysteresis properties of fine grain titanomagnetites (~ 0.1 microns).

Composition (x)	H_c (Oe)	H_{rc} (Oe)	$\dfrac{H_{rc}}{H_c}$	$\dfrac{J_{rs}(H=0)}{J_s(H=\infty)}$	x_0 (G·Oe^{-1})
0	396	565	1.43	0.380	0.149
0.1	495	690	1.39	0.429	0.124
0.2	660	940	1.42	0.473	0.096
0.3	693	930	1.34	0.500	0.084
0.35	1,171	1,520	1.30	0.594	0.053
0.5	1,410	1,690	1.20	0.564	0.039
0.55	1,860	2,200	1.18	0.583	0.030
0.60	1,910	2,140	1.12	0.474	0.029

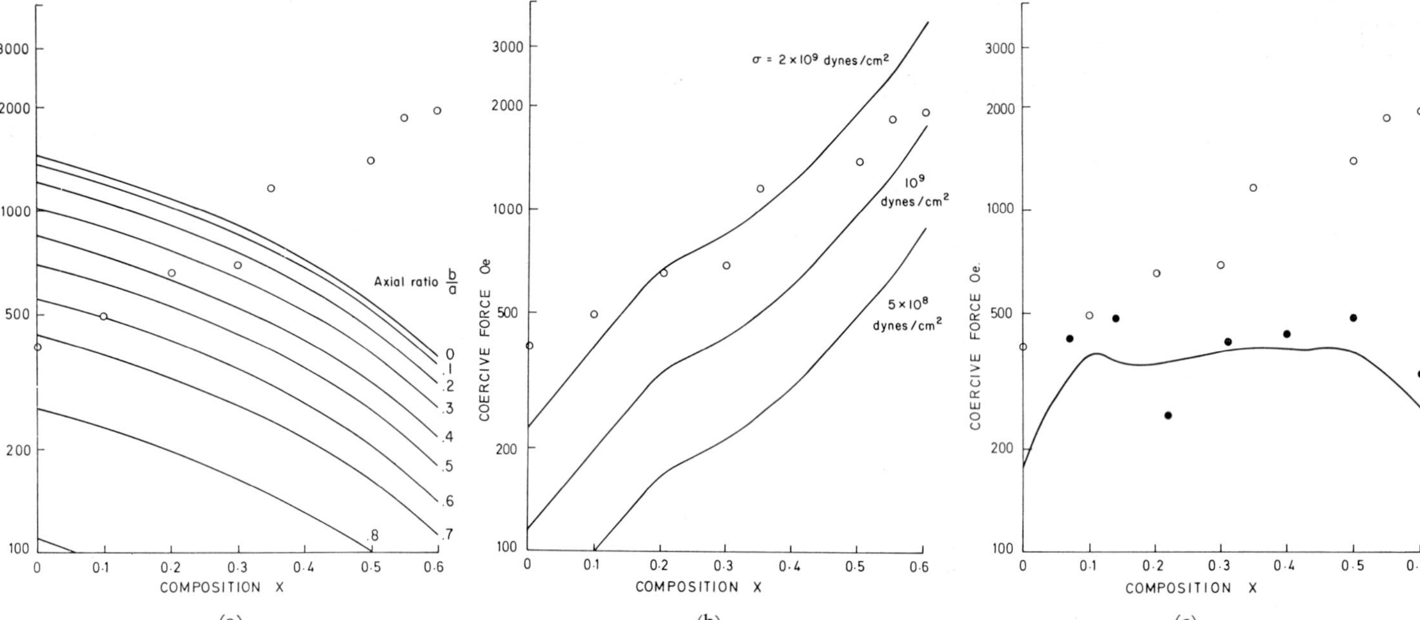

Fig. 2. Variation of coercive force with composition in fine grain titanomagnetite. Open circles represent SD samples (Table 2): closed circles, values from ROBINS (1972) with $d<1.5m$; solid lines, theoretical values calculated by using the Stoner-Wohlfarth model (magnetic parameters taken from SYONO (1965). (a) Shape anisotropy is $H_c=0.479\,(N_b-N_a)J_s$, (b) stress anisotropy is $H_c=0.479\,(3\lambda\sigma/J_s)$, and (c) magnetocrystalline anisotropy is $H_c=0.64K_1/J_s$.

Table 3. Hysteresis properties of coarse grain titanomagnetite (>150 microns).

x	H_c (Oe)	H_{rc} (Oe)	$\dfrac{H_{rc}}{H_c}$	$\dfrac{J_{rs}(H=0)}{J_s(H=\infty)}$	x_0 (G·Oe^{-1})
0.0	31	162	5.22	0.018	0.212
0.1	30	133	4.33	0.028	0.350
0.2	27	118	4.37	0.032	0.304
0.3	36	131	3.64	0.045	0.187
0.4	31	115	3.71	0.039	0.242
0.5	36	133	3.65	0.049	0.257
0.6	33	129	3.91	0.052	0.268
Mean	32	129	4.14	—	0.260

Table 4. Hysteresis properties of sized titanomagnetites ($x=0.4$).

Size (microns)	Standard deviation	H_c (Oe)	H_{rc} (Oe)	H_{rc}/H_c	J_{rs}/J_s	x_0 (G·Oe^{-1})
0.8	—	1,551	1,900	1.23	0.449	0.036
0.95	—	1,076	1,600	1.49	0.427	0.054
1.64	0.9	594	1,030	1.73	0.362	0.088
3.5	1.8	275	580	2.11	0.250	0.137
6.0	2.6	152	385	2.53	0.146	0.182
8.6	2.9	112	320	2.86	0.095	0.208
13.0	3.2	82	270	3.29	0.079	0.224
17.0	5.1	66	235	3.56	0.053	0.236
32	8.7	44	210	4.77	0.034	0.265
61	18	39	185	4.74	0.030	0.279
84	16	36	176	4.89	0.027	0.294
112	21	34	186	4.94	0.033	0.271

theoretical minimum of 1.09 ($1.43 > H_{rc}/H_c > 1.12$). The magnetic hardness (a hard magnetic material has a large coercive force) increases with increasing titanium content (Table 2). The large observed coercivities in the titanium rich titanomagnetites can only be due to magnetostrictive anisotropy. This is clearly demonstrated in Fig. 2, where the observed coercive force is compared to the theoretical values calculated from the STONER and WOHLFARTH (1948) SD theory. MD titanomagnetites (Table 3) show no systematic variation of magnetic properties with composition, suggesting that inclusions are the dominant cause of the coercivity. However, no additional evidence is available at present to substantiate this.

The effect of particle size on magnetic properties has been investigated in natural magnetites (PARRY, 1965; RAHMAN et al., 1973; DUNLOP, 1973a; LEVI, 1974; JOHNSON et al., 1975; LEVI and MERRILL, 1977; BAILEY, 1975) and synthetic titanomagnetites (ROBINS, 1972; DAY, 1973; DAY et al., 1977).

Table 4 shows the variations observed in an intermediate titanomagnetite. The trends seen in the table are typical of the titanomagnetites, and many other magnetic materials (see LUBORSKY, 1961).

The coercive force, H_c, attains a maximum at, or just below, the SD transition

size. The data available from titanomagnetites does not show the maximum, but it
has been observed in iron (Luborsky, 1961) and other ferrites (Berkowitz and
Schuele, 1959) and it is expected on theoretical grounds as a result of the onset of
superparamagnetism. Above the SD transition, the coercivity decreases continuously
with increasing particle size following a power law $H_c \propto d^{-n}$. Values of n have been
given in the range $0.4 < n < 0.9$. The smaller values are for annealed natural magnetite (Parry, 1965), while the larger values are for synthetic (are probably stressed)
titanomagnetites (Day, 1973). Stacey and Wise (1967), Dunlop (1976), and Day
et al. (1977) have discussed the theoretical basis for the power law dependence.

Above about 20–30 μm the index, n, is much lower, usually between 0 and 0.3.
This transition size usually coincides with the critical size for the change from PSD
to true MD behaviour.

Remanent coercivity, H_{rc} behaves the same way, but with power indices less
than those found for H_c. The interpretation of the variation in isothermal remanence, J_r, and susceptibility is not so readily treated as the coercivity because the
data shows much more scatter.

The original size distribution of magnetic carriers can be modified by a number
of mechanisms.

1) Sub-microscopic inclusions in silicates (e.g., Evans and Wayman, 1970;
Evans, 1977, this issue).

2) Subdivision by exsolution lamellae (e.g., Larson et al., 1969; Evans, 1977,
this issue).

3) Stress controlled SD regions (e.g., Verhoogen, 1959; Shive, 1969a, b).
In addition, the original distribution may be such that minor amounts of fine SD
particles are present along with a greater proportion of larger MD particles. Also,
there exists the possibility that a single sample may possess both a soft magnetic

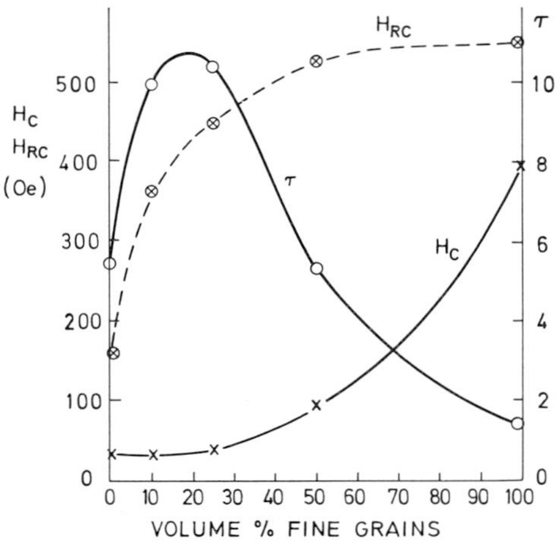

Fig. 3. Coercive force H_c, remanent coercive force H_{rc},
and the ratio H_{rc}/H_c for mixed samples (see text),
plotted against volume percentage of fine grain fraction.

fraction and a hard magnetic fraction (DUNLOP, 1969). In all these cases, the effect of mixtures must be considered. MEIKELJOHN (1953), KNELLER and LUBORSKY (1963), and WASILEWSKI (1973) have investigated the magnetic properties of mixtures. To examine this effect in extreme cases DAY *et al.* (1976) prepared a set of samples containing two size fractions in various proportions. In Fig. 3, the soft component has $H_c = 31$ Oe, $H_{rc} = 162$ Oe and the hard component has $H_c = 396$ Oe and $H_{rc} = 565$. For samples containing less than 20 per cent of the hard fraction, the coercive force is almost constant and is dominated by the coercive force of the soft fraction, while the remanent coercivity increases rapidly. The opposite is true in the samples that contain 50 per cent or more of the hard fraction where the remanent coercivity remains constant and close to the value of the hard fraction, while the coercive force increases. These experiments show that H_{rc} or H_c alone may not always indicate the true stability. However, in paleomagnetism, where AF demagnetization is used, the whole spectrum of coercivities is examined (see LARSON *et al.*, 1969).

Additional factors that can affect magnetic properties are residual stress (DAY, 1973) and grain shape and grain surface irregularities (BANERJEE, 1977, this issue).

4.   Theoretical Models of TRM

4.1   *SD TRM*

NÉEL (1949, 1955) proposed a model of TRM in single-domain grains that follows directly from his discussion of superparamagnetism. His starting point was to consider a noninteracting dispersion of SD particles with uniaxial anisotropy, K_u, volume, V, and saturation magnetization, J_s. The IRM induced by a weak field, H, would be small because only those particles whose critical field $(2K_u/J_s)$ was smaller than H would be magnetized. Further, this IRM could be destroyed by an alternating field equal to the inducing field. However, if the dispersion is heated the IRM acquired at a given temperature increases because the critical field decreases. Just below the Curie point the critical field vanishes so that all particles could be magnetized by a vanishingly small field. However, this mechanism produces a TRM that is equal to J_s and also independent of the inducing field. To overcome these difficulties Néel introduced thermal fluctuations. The basic equation of thermal activation gives the probability, dp, of a change occurring in a short time, dt, in terms of the activation energy, E, and the absolute temperature, T:

$$dp = C \exp\left[-E/kT\right]dt .$$

This leads to a characteristic time for the process, which appears as the relaxation time, τ, in theories of TRM:

$$1/\tau = dp/dt = C \exp\left[-E/kT\right] .$$

A single-domain particle with uniaxial anisotropy possesses two stable positions separated by a potential barrier, $E\,(=K_u V,$ in the absence of a field). In the absence of an applied field, both stable positions have equal probabilities for occupation and a remanent magnetization of initial value $J_R(0)$ will decay:

$$J_R(t) = J_R(0) \exp\left[-t/\tau\right]$$

$$1/\tau = f_0 \exp\left[-K_u V/kT\right]$$

where f_0 (which has replaced C) is called the frequency factor. NÉEL (1949) investigated the nature of the perturbing couples which determine f_0. Deformations of the grain by thermal fluctuations cause a change in both the magneto-elastic energy and the demagnetizing coefficient. These deformations give rise to a total couple

$$r = |3g\lambda + NJ_s^2|(2V/\pi gkT)^{1/2}$$

giving the frequency factor, f_0, as

$$f_0 = \frac{2eK_u}{mJ_s}|3g\lambda + NJ_s^2|(2V/\pi gkT)^{1/2}$$

where g is the shear modulus, λ the magnetostriction coefficient, N the demagnetization coefficient, and e is the charge and m the mass of the electron. BROWN (1959) criticized Néel's derivation of f_0, but his calculation gives a value of f_0 that does not differ significantly from Néel's value (10^{10} sec^{-1}).

The relaxation time is a strong function of temperature so that a relatively small change in temperature gives rise to a sharp increase in the relaxation time. This leads immediately to Néel's concept of a blocking temperature. In the presence of an applied field the stable positions are split into two energy levels so that the occupation probability of the level that corresponds to a magnetization parallel to the applied field is larger than that antiparallel. Néel showed that the magnetization at the blocking temperature was then given by:

$$J_{T_B} = VJ_s(T_B) \tanh \frac{VHJ_s(T_B)}{kT_B} \, .$$

The direction of the magnetization is fixed at the blocking temperature. On cooling the relaxation time increased sharply and the acquired TRM increases only as J_s changes with falling temperature:

$$\langle J_{\text{TRM}} \rangle = VJ_s(T_0) \tanh \frac{VHJ_s(T_B)}{kT_B} \, .$$

For an assemblage of n aligned grains the intensity is increased by a factor n. Néel made the assumption that a random assemblage can be replaced by an ideal one where one-third of the grains have easy axes parallel to the applied field. STACEY and BANERJEE (1974), however, assumed a random orientation, then

$$J_{\text{TRM}} = J_s(T_0) \int_0^{\pi/2} \tanh \frac{VJ_s(T_B)H \cos\theta}{kT_B} \cos\theta \sin\theta \, d\theta \, .$$

The general form of this function has been tabulated by STACEY and BANERJEE (1974, Appendix 1).

Néel's model is not readily amenable to precise quantitative checks because it requires grains in a limited size range and with constant shape so that V and T_B is the same for all of them [although DUNLOP and WEST (1969) and DUNLOP (1976) have had some success by determining the grain size-coercivity distribution of each sample experimentally]. However, it is remarkably successful in explaining the following characteristics of TRM:

1) It defines (via the relaxation time being equal to some measurement time) a blocking temperature.

2) It accounts for the law of additivity of PTRM's.

3) TRM intensity is much larger than that of an IRM acquired in the same small field.

4) TRM is very stable to
    a. AF demagnetization because H_c at room temperature is much greater than H_c at the blocking temperature.
    b. Thermal demagnetization because T_B is usually well above room temperature.
    c. Time because of the exponential dependence of the relaxation time.

4.2   MD TRM

In multi-domain grains, remanent magnetization is induced by irreversible domain wall movements past energy barriers. Since this process can be thermally activated the domain walls can have blocking temperatures. However, the blocking process is not as simple as with single-domains because all the domain walls in a single grain lock into a pattern and mutually interact, i.e., TRM is a property of the whole grain.

4.2.1   Néel MD model

NÉEL (1955) approached MD TRM via the expansion and contraction of a rectangular hysteresis loop with changing temperature. Both J_s and H_c are temperature dependent. Néel assumed:

$$J_s \propto (T_c - T)^{1/2} \text{ near } T_c$$
$$H_c \propto (T_c - T)$$

hence

$$\frac{H_c(T)}{H_c(T_0)} = \left[\frac{J_s(T)}{J_s(T_0)}\right]^2 .$$

At T_c, the hysteresis loop is infinitely small. As the temperature drops the loop expands, with H_c growing faster than J_s. In effect, H (applied field) moves in with respect to the loop (Fig. 4). The hysteresis loop is "sheared" by the internal demagnetizing field, hence at some temperature "H will start to descend the back slope" (Fig. 4). When $H = H_c(T)$ the magnetization becomes fixed; it cannot ascend BC because this would require a minor loop. Quantitatively the magnetization becomes blocked when the ratio AQ/AB (Fig. 4c) is a minimum (note that this ratio is just $J(T)/J_s(T)$).

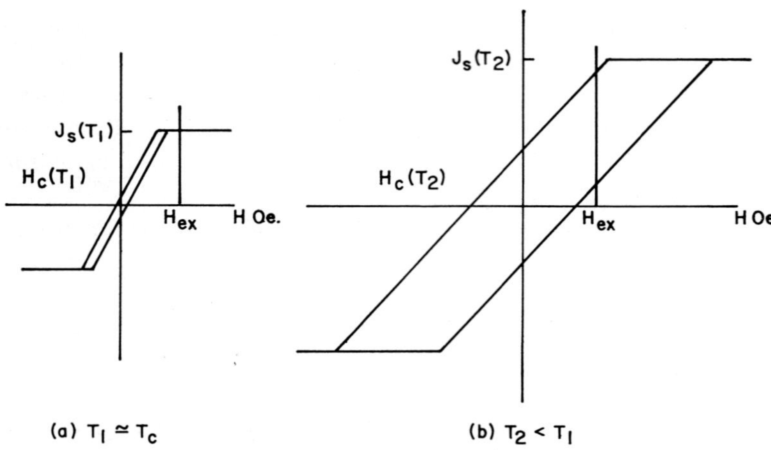

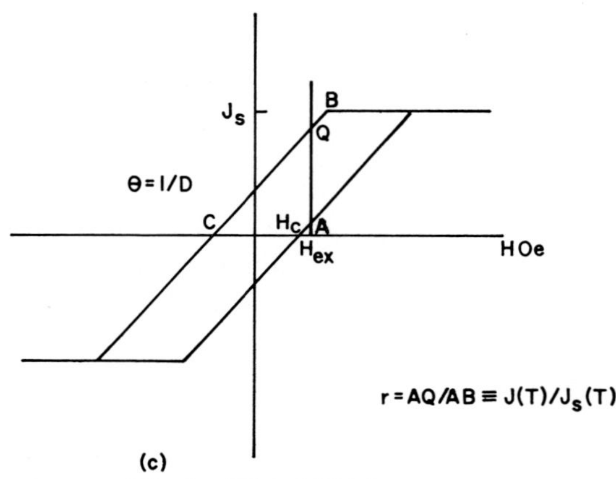

Fig. 4.   Néels MD Theory.

From Fig. 4c

$$\frac{AQ}{AB} = r = \frac{H_c(T) + H}{N J_s(T)}$$

for minimum

$$\frac{dr}{dT} = 0 = \frac{1}{N J_s(T)} \cdot \frac{dH_c(T)}{dT} - \frac{H_c(T) + H}{N J_s^2(T)} \cdot \frac{dJ_s(T)}{dT}$$

hence,

$$\frac{dH_c(T)}{dT} = \frac{H_c(T) + H}{J_s(T)} \cdot \frac{dJ_s(T)}{dT}$$

from the dependence of $H_c(T)$ on $J_s(T)$, we get

$$\frac{dH_c(T)}{dT} = 2H_c(0)\frac{J_s(T)}{J_s^2(0)} \cdot \frac{dJ_s(T)}{dT} .$$

Eliminating $dH_c(T)/dT$, we get the blocking condition

$$H = H_c(T_B) .$$

At $r_{\min}$, $T = T_B$, hence:

$$r(T_B) = \frac{2H}{N J_s(T_B)}$$

therefore,

$$\begin{aligned}
J_{\mathrm{TRM}} &= r(T_B) J_s(0) \\
&= \frac{2H J_s(0)}{N J_s(T_B)} \\
&= 2\frac{H}{N}\left[\frac{H_c(0)}{H_c(T_B)}\right]^{1/2} \\
&= \frac{2}{N} H^{1/2} \cdot H_c^{1/2}(0) .
\end{aligned}$$

For low fields ($\sim$ few Oe), Néel invoked a thermal fluctuation field, H_f, such that a wall cannot be blocked until $H_c(T_B) = H_f$. This is the blocking condition for $H < H_f$. When $H_c(T_B) = H_f$,

$$\frac{H_f}{H_c(T_B)} = \left[\frac{J_s(T_B)}{J_s(0)}\right]^2 .$$

The magnetization at T_B is simply H/N; therefore, at room temperature

$$J_{\mathrm{TRM}} = \frac{H}{N} \cdot \frac{J_s(0)}{J_s(T_B)} = \frac{H}{N} \cdot H_f^{1/2} \cdot H_c^{1/2}(0) .$$

The transition between the linear field dependence and the square root dependence occurs at $H = 4H_f$ and the TRM is saturated for $H > (1/4)H_c(0)$.

4.2.2 Stacey MD model

STACEY (1958) suggested that at high temperatures ($T_c > T > T_B$), the minimum energy configuration simply corresponds to zero internal field. He argues that only the magnetostatic energy and the interaction energy with the applied field need be considered.

The magnetic energy above T_B is

$$E = -HJ + 1/2 N J^2$$

for minimum

$$\frac{\mathrm{d}E}{\mathrm{d}J} = 0 = -H + NJ .$$

This, of course, is equivalent to saying that the internal field is zero. Above T_B, the magnetization is given by $J = H/N$ (cf., Néel, previous page). At T_B, the magnetization is blocked and the magnetization is fixed except for modifications in the intensity due to an increase in J_s on cooling and the effect of the demagnetizing field. At room temperature, the TRM is given by

$$J_{\mathrm{TRM}} = \frac{H}{N} \cdot \frac{1}{1 + Nx_0} \cdot \frac{J_s(0)}{J_s(T_B)}$$

where x_0 is the initial susceptibility.

Notice that Stacey's model does not say anything about the blocking process nor does it derive a blocking condition explicitly. Later, we will see that this model can be regarded as an extreme case of a generalized Néel model.

4.2.3 Everitt model

EVERITT (1962b), noting that the probability of a wall traversing a particular energy barrier may increase with increasing temperature through an increase in thermal agitation or through a decrease in the magnitude of the barrier height, proposed a relaxation time for domain wall movement as

$$\frac{1}{\tau} = \frac{1}{\tau_0} \exp\left[-E/kT\right]$$

where E is the barrier height and τ_0 is related to thermal agitation. NÉEL (1955) derived the relation:

$$2J_s H_c = \frac{1}{S}\left(\frac{\mathrm{d}E}{\mathrm{d}X}\right)_{\max}$$

when S is the surface area of the wall and X is a measure of the displacement of the wall. Everitt approximates

$$\left(\frac{\mathrm{d}E}{\mathrm{d}X}\right)_{\max} \simeq \frac{E}{c\lambda}$$

where λ is the barrier width and c is a factor determined by the shape of the barrier, giving

$$\frac{1}{\tau} = \frac{1}{\tau_0} \exp\left(-\frac{2cS\lambda H_c J_s}{kT}\right).$$

When the temperature is raised, kT increases while H_c and J_s decrease. Therefore, there should always be a temperature, T_B, below T_c, where τ becomes comparable to the duration of an experimental measurement. Once again the exponential term causes T_B to be sharply defined. Setting $\tau = \tau_B$ and taking logarithms we obtain:

$$\frac{H_c(T_B)J_s(T_B)}{T_B} = \frac{k \log \tau_B/\tau_0}{cS\lambda}.$$

Everitt further assumes the temperature dependences of J_s and H_c to be of the form

$$\frac{H_c(T)}{H_c(0)} = \left[\frac{T_c - T}{T_c}\right]^l$$

$$\frac{J_s(T)}{J_s(0)} = \left[\frac{T_c - T}{T_c}\right]^m$$

giving the blocking relations above as

$$\frac{1}{T_B}\left[\frac{T_c - T_B}{T_B}\right]^{l+m} = \frac{k \log \tau_B/\tau_0}{cS\lambda H_c(0)J_s(0)}.$$

When a small magnetic field, H, is applied at an angle, α, to the wall the barrier height is modified to:

$$E' = \lambda S J_s (c H_c(T) - H \sin \alpha)$$

giving a complicated form of the blocking condition

$$\frac{1}{T_B}\left\{ \left(\frac{T_c - T_B}{T_c}\right)^{l+m} - \frac{H \sin \alpha}{c H_c(0)}\left(\frac{T_c - T_B}{T_c}\right)^m \right\} = \frac{k \log \tau_B/\tau_0}{c S \lambda H_c(0) J_s(0)}$$

and the intensity of TRM as

$$J_{\mathrm{TRM}} = \frac{H \sin \alpha}{N}\left(\frac{H_c(0)}{a T_B}\right)^{m/l+m}\left\{ 1 - \frac{m}{l+m}\frac{H \sin \alpha}{H_c(0)}\left(\frac{H_c(0)}{a T_B}\right)^{l/l+m} \right\}$$

where

$$a = \frac{k \log \tau_B/\tau_0}{c S \lambda H_c(0) J_s(0)} \, .$$

In higher fields, when thermal agitation is not important, Everitt generalized Néel's MD model by using

$$\frac{J_s(T)}{J_s(0)} = \left[\frac{H_c(T)}{H_c(0)}\right]^{m/l} \, .$$

This differs from Néel's equation only by the general index rather than the particular value of 2.

4.2.4   Schmidt MD model

The Schmidt model (SCHMIDT, 1973) considers the temperature variation of the total energy of an ideal rectangular 2 domain grain. The total energy has contributions due to interactions between the domain wall and (1) the external field, (2) the demagnetizing field, and (3) a regular array of lattice defects. The total energy expression is therefore

$$E = -AHm(T)X + Dm^2(T)X^2 - Bm^p(T)\cos\left(\frac{2\pi X}{l}\right)$$

where the notation is given in Fig. 5.

The energy can be minimized to give the maximum stable wall position, W. The blocking condition is then given by minimizing W with respect to the reduced magnetization $m(T)$ (which contains the temperature variation in the model), giving

$$W_{\min} = \frac{\pi B m_B^{P-2}}{D} + \frac{AH}{2 D m_B}$$

$$m_B^{P-1} = \frac{AlH}{2(P-2)\pi B}$$

The magnetization J_{T_B}, is $A m_B W_{\min}$ so that the intensity of TRM is given by

$$J_{\mathrm{TRM}} = m_B W_{\min}\frac{m(0)}{m_B}$$

hence,

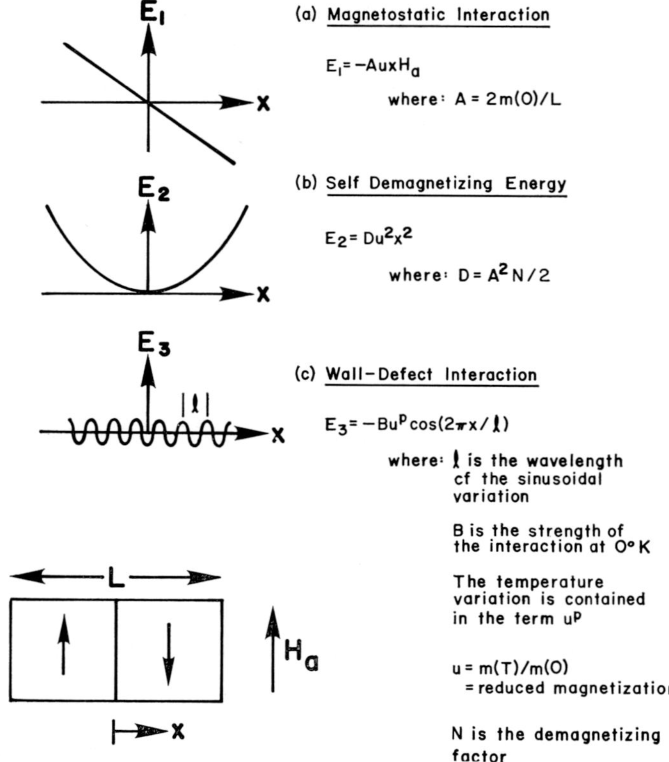

Fig. 5.   Notation for Schmidt's MD Theory.

$$J_{\mathrm{TRM}} = \frac{A^2(P-1)}{2D(P-2)} \cdot \left[\frac{2(P-2)\pi B}{Al}\right]^{1/P-1} \cdot m(0) \cdot H^{(1-(1/P-1))} .$$

The coercive force, $H_c(T)$, calculated from the model, is

$$H_c = \frac{2\pi Bm}{Al}^{P-1}$$

so that the intensity of TRM can finally be written as

$$J_{\mathrm{TRM}} = \frac{A^2(P-1)}{2D(P-2)} \cdot \{(P-2)H_c(0)\}^{1/P-1}H^{(1-(1/P-1))} .$$

Again, in low fields thermal activation must be considered. Schmidt assumes that "thermal blocking" is governed by the equation,

$$dp = C \exp(E/kT)dt \quad [\text{cf., SD equation}]$$

and defines the blocking temperature, T_{ThB}, as the temperature at which, for $dp = e^{-1}$, dt is consistent with the cooling rate of the magnetic material. Numerical solutions yield a linear dependence of TRM on applied field, H, and a blocking temperature that is independent of field. The field below which thermal activation blocking is dominant is calculated to be

$$H_T = H_c(0) \cdot (P-2)\left(\frac{P-1}{P-2}\right)^{P-1}\left(\frac{m_{\mathrm{THB}}}{m(0)}\right)^{P-1}.$$

4.3 Comparison of MD models

A closer examination of these models shows that they are in fact variations of the same model. DUNLOP and WADDINGTON (1975) generalized Néel's model by retaining a general value of n (Néel chose $n=2$) in the relation

$$H_c(T) = \beta J_s^2(T)$$

(This, of course, is essentially the same as the treatment by Everitt.) The generalized model yields the blocking condition

$$H_c(T_B) = \frac{1}{n-1}H$$

and an intensity of TRM

$$J_{\mathrm{TRM}} = \frac{n}{(n-1)^{1-(1/n)}} \cdot \frac{1}{N} \cdot H^{1-(1/n)} \cdot H_c^{1/n}.$$

Dunlop and Waddington went on to show that the extreme cases of $n=2$ and n large yield the Néel and Stacey MD models, respectively. Further, it can be shown that both the Schmidt model and the Everitt model give the same blocking and induction equations if we relate the indices m, 1, p and n by

$$n = p-1 = \frac{l}{m}.$$

For low fields, again it can be shown that the models are physically equivalent for further discussion see DUNLOP and WADDINGTON, 1975). In the generalized Néel model,

$$J_{\mathrm{TRM}} = \frac{H}{N}\left[\frac{H_c(0)}{H_f(T_B)}\right]^{1/n}$$

and the breakpoint between H and $H^{1-(1/n)}$ dependences occurs at a field:

$$H_t = \frac{n^n}{(n-1)^{n-1}}H_f(T_B)$$

MERRILL (1977, this issue) examines the nature of the demagnetizing factor, N, which appears in all MD TRM models. He shows that, in the limit of very small domain wall displacements, the demagnetization factor can be taken as a constant, independent of wall displacement. This is the assumption used in the MD TRM models, but Merrill's calculation shows that "the factor N is related to domain structure and not to the shape of the grain. The grain's shape will enter only to the extent that it affects the domain structure." Merrill further suggests that N is a monotonically increasing function of the mean distance of the domain wall displacement from the zero field equilibrium configuration. This implies that the acquisition of TRM will not vary simply as a function of the external field raised to some power.

4.4 PSD models

Much of the recent theoretical work in rock magnetism has been concerned with the magnetic properties of PSD grains, particularly TRM. A detailed discussion of this work can be found elsewhere in this issue (DUNLOP, 1977; BANERJEE, 1977). Briefly, four mechanisms have been proposed whereby fine grains (above the SD threshold) may possess a PSD moment. These are:

   1)   Barkhausen discreteness of domain wall movement (STACEY, 1962).

   2)   Domain wall moments, particularly 'Psarks' (DUNLOP, 1977).

   3)   Moments pinned by stress fields of dislocations (VERHOOGEN, 1959; OZIMA and OZIMA, 1965; KOBAYASHI and FULLER, 1968; SHIVE, 1969a, b).

   4)   Surface moments (STACEY and BANERJEE, 1974; BANERJEE, 1977).

The permanent moment implies that the TRM induction is SD like. For mechanisms 1, 3 and 4, the intensity of TRM varies inversely as the grain diameter.

5. Particle Size and TRM

5.1 TRM induction

Figure 6 shows typical induction curves for sized titanomagnetites ($X=0.4$ and 0.6). These compositions were chosen because their Curie temperatures are well below the temperatures where annealing and chemical and textural alteration take place. As a check, the saturation IRM and its AF demagnetization curve was

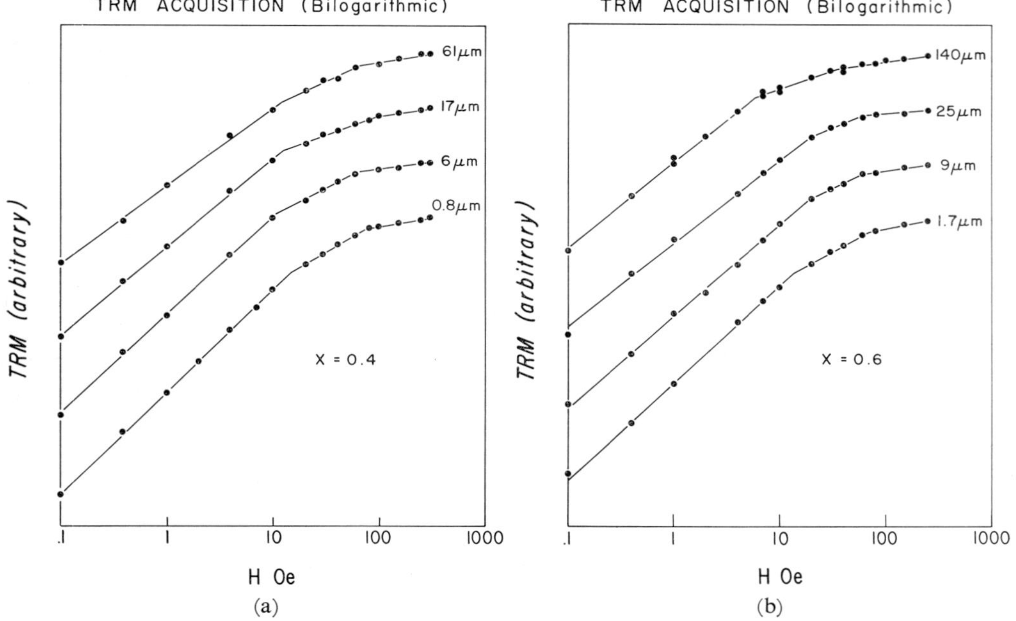

Fig. 6. TRM induction curves: (a) Titanomagnetite $X=0.4$. (b) Titanomagnetite $X=0.6$. (Sample numbers on curve correspond to those in Table 5.)

Table 5.   Hysteresis properties of the samples in Fig. 6.

	Composi-tion (X)	Size (microns)	H_c (Oe)	H_{rc} (Oe)	H_{rc}/H_c	J_{rs}/J_s	x_0 (G/Oe)	Domain state
1	0.4	$\lesssim 0.8$	1,551	1,900	1.25	0.449	0.036	SD
2	0.4	6 ± 3	152	385	2.53	0.146	0.186	PSD
3	0.4	17 ± 5	66	235	3.56	0.053	0.236	PSD/MD
4	0.4	61 ± 18	39	185	4.74	0.030	0.279	MD
1	0.6	1.7 ± 1	581	1,115	1.94	0.459	0.048	SD/PSD
2	0.6	9 ± 3	92	260	2.83	0.131	0.118	PSD
3	0.6	25 ± 8	61	185	3.03	0.063	0.237	PSD/MD
4	0.6	140 ± 24	39	148	3.79	0.043	0.268	MD

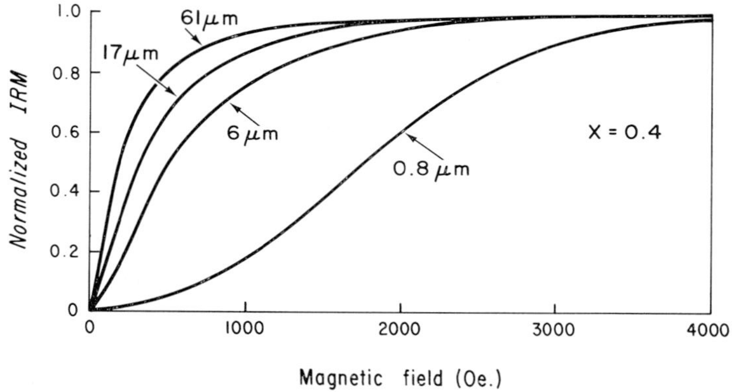

Fig. 7.   IRM acquisition for titanomagnetite ($X=0.4$).

measured after each heating, and the TRM induced in some given field was reproduced at the end of the experiment.

The samples were chosen so as to span the SD, PSD and MD size ranges (Table
5).   However, the induction curves show no strong variation with either domain
state or particle size.   Contrast this with the IRM induction curves (Fig. 7), where
there is a strong dependence on size.   The bilogarithmic plots of the TRM induction
curves show three linear segments (Table 6).   In weak fields, the slope increases for
decreasing particle size, and is close to 1 for the smallest sizes.   Above about 15 Oe,
the field dependence is weaker, the slope is about one-half of the weak field slope,
and again the slope seems to increase for decreasing particle size.   The third region,
above about 70 Oe, has an even weaker field dependence.   The slope is less than 0.2.

This type of curve is not restricted to these particular samples; the same behaviour is exhibited by sized synthetic magnetites (DUNLOP, 1973a), cobalt ferrite
and a variety of rock types (DUNLOP and WADDINGTON, 1975).   In fact, the author
has not seen a *detailed* TRM induction curve that does not look like those shown in
Fig. 6.

Table 6. TRM data.

Sample		Grain size (microns)	Coercive force (Oe)	n_1	n_2	n_3	H_{12} (Oe)	H_{23} (Oe)	n_{calc}	n	Q_t
Titanomagnetites											
$X=0.4$	1	0.8	1,551	0.99	0.59	0.185	15	76	2.44	—	51
	2	6	152	0.95	0.53	0.14	11.5	63	2.13	—	5.2
	3	17	66	0.86	0.40	0.18	12.5	85	1.67	—	2.0
	4	61	39	0.77	0.49	0.185	12.5	66	1.96	—	1.1
$X=0.4$		1.6	594	0.93	0.55	0.27	11.3	60	2.22	—	20
"		3.5	275	0.92	0.54	0.15	16	64	2.17	—	12
"		8.6	112	0.91	0.53	0.165	10	62	2.13	—	3.7
"		13	82	0.84	0.46	0.18	12	70	1.85	—	2.7
"		32	44	0.82	0.42	0.12	12	74	1.72	—	1.3
"		84	36	0.79	0.41	0.10	13.5	77	1.69	—	0.9
"		110	34	0.71	0.34	0.18	15	82	1.52	—	1.6
$X=0.6$	1	1.7	581	0.95	0.57	0.20	13.5	67	2.33	—	29
	2	9	92	0.90	0.50	0.15	19	60	2.0	—	4.6
	3	25	61	0.81	0.43	0.08	20	60	1.75	—	1.4
	4	140	39	0.83	0.38	0.16	6	36	1.61	—	1.6
$X=0.6$		0.8	1,584	0.96	0.58	0.30	15	59	2.38	—	69
"		3.3	280	0.93	0.62	0.15	13.5	60	2.63	—	15
"		6.4	132	0.96	0.65	0.12	14.5	54	2.86	—	6.3
"		12	82	0.96	0.58	0.19	16	40	2.38	—	2.5
"		16	73	0.87	0.52	0.11	19	55	2.08	—	1.5
"		41	52	0.69	0.39	0.07	10	70	1.64	—	1.6
"		78	46	0.76	0.33	0.14	9.5	97	1.49	—	1.1
Fe_3O_4	1	0.037	205	0.76	0.38	—	11	—	1.75	1.96	26
	2	0.076	155	0.37	0.43	—	7	—	1.54	1.82	44
	3	0.1	137	0.52	0.30	—	6.5	—	1.43	1.79	45
	4	0.22	92	0.48	0.28	—	6.5	—	1.39	1.47	27
$CoFe_3O_4$		$\sim$0.1	1,025	0.83	0.54	—	4.5	—	2.17	—	—
$\gamma Co_{0.03}Fe_{1.97}O_3$		$\sim$0.05	875	0.85	0.53	—	4.5	—	2.13	—	—
Basalt		—	300	0.84	0.47	—	9	—	1.89	1.32	15.5
Basalt		—	254	0.96	0.49	—	30	—	1.96	1.61	11
Basalt		—	176	0.88	0.52	0.23	18	160	2.08	1.79	6.1
Diabase		—	106	0.61	0.22	0.09	19	90	1.28	1.08	8.5
Dolerite		—	96	0.90	0.66	—	14	—	2.94	1.39	3.9
Gabbro		—	54	0.68	0.37	—	9	—	1.59	1.61	5.8
Anorthosite		—	38	0.60	0.49	—	10	—	1.96	1.12	3.6
Granite		—	31	0.92	0.50	0.145	20	140	2.0	1.64	1.0
Gabbro		—	26	0.67	0.39	—	15	—	1.64	1.22	1.4
Diorite		—	12	0.71	0.43	—	9	—	1.75	1.25	0.9

$n_1 n_2 n_3$ are slopes of TRM induction curves (Fig. 6).

$H_{12} H_{23}$ are transition fields in TRM induction curves (Fig. 6).

n is the index in $H_c \propto J_s^n$.

n_{calc} is obtained from $n=(1/1-n_1)$ (see Section 4.2).

(Data from DAY, 1973; DUNLOP, 1973a; DUNLOP and WADDINGTON, 1975).

The TRM induction curve can, therefore, be characterized on a bilogarithmic plot by three linear segments and two transition fields. These parameters are given in Table 6. The overall shape of these curves are not unlike that predicted by the MD TRM models. However, as we shall see later, there are many discrepancies between the observed curves and TRM theory.

5.2 *TRM intensity*

Figures 8 and 9 show how specific TRM (TRM/unit field) varies with grain size. The scatter in the data is not surprising since the data has been compiled from many laboratories and contains results from natural and synthetic powders. The curves, however, do have some similar features. At low sizes there is an indication of a peak in the TRM curve followed in the intermediate size range by a definite size dependence. For the larger grains, the grain size dependence is less marked. Earlier

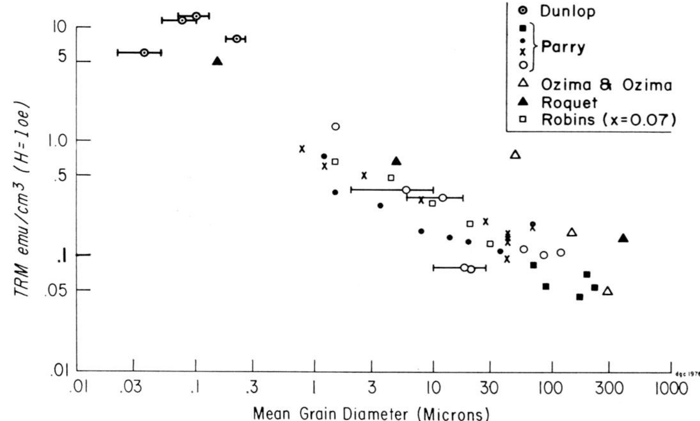

Fig. 8. Weak field specific TRM in magnetite as a function of grain size.

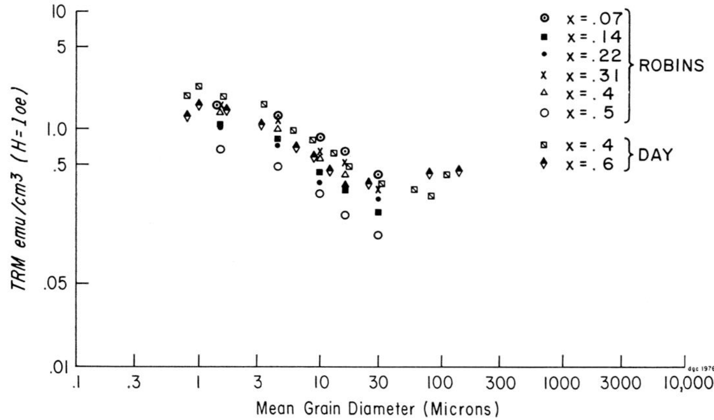

Fig. 9. Weak field specific TRM in titanomagnetite (X=0) as a function of grain size.

curves of this type (e.g., STACEY and BANERJEE, 1974) were used to show an exact inverse dependence of specific TRM on grain size (a property of PSD TRM theory), but the present compilation of data indicates that a $1/d$ dependence is not quite correct. The variation

$$\frac{J_{\mathrm{TRM}}}{H} \propto d^{-l}$$

provides a better fit to the intermediate data when l is about 0.7. The dependence on grain size in the larger grains again differs from previous results where it was concluded that MD TRM was insensitive to variations to grain size (PARRY, 1965; DICKSON et al., 1966). The intensity of TRM in large single crystals (LEVI, 1974), whose size is too large to show on Fig. 8, is approximately 0.008 emu cm^{-3}, showing that the intensity of TRM decreases with grain size over a considerable size range.

The Koenigsberger ratio, Q_t, decreases very rapidly with increasing particle size (Table 6) from 60–70 for the finest grains to about 1 for the coarse grains.

5.3   AF demagnetization

AF demagnetization curves of TRM and saturation IRM are dependent on particle size (RIMBERT, 1959). ROQUET (1954), RIMBERT (1959) and DUNLOP and WEST (1969) investigated the stability trends of saturation IRM, TRM, and ARM, and showed that for SD particles both the TRM and ARM stability decreased as the inducing field increased. The stability of TRM was always greater than that of the saturation IRM. LOWRIE and FULLER (1971) investigated the stability of TRM in multi-domain particles and found that (1) the stability to AF demagnetization increased for increasing inducing fields, and (2) the stability of TRM was less than that of the saturation IRM. On the basis of their own observations and the previous work on SD particles LOWRIE and FULLER (1971) suggested that these stability trends could be used to distinguish between the SD and MD states. Although the so-called Lowrie-Fuller test is used extensively in paleomagnetism, its physical basis is not clear and there is some doubt as to what information the test yields, or even if the test is applicable. The main problem is the nature of the transition in behaviour. Does it occur at the SD-PSD transition or the PSD-MD transition, or is the phenomenon more complex? Recent measurements indicate that it is complex. Preisach analysis of the AF demagnetization curves (BAILEY, 1975; BAILEY et al., 1975) indicates that the Lowrie-Fuller test 'does not respond to the onset of PSD moments, but it nevertheless accurately delineates the PSD region over which SD and MD characteristics are mixed.' The Preisach analysis also yields further information on the physical basis of the test; The Lowrie-Fuller test gives a MD result for any coercivity spectrum more concave than an exponential function and, hence, the test distinguishes between the peaked coercivity spectrum of SD particles and the sub-exponential MD spectrum. DUNLOP (1973a) suggests that the shapes of the AF demagnetization curve (which is, of course, the coercivity spectrum) can be used to distinguish between SD and MD particles for the same reasons as were outlined above.

Many examples have been found where the stability trends change with increasing demagnetizing field, i.e., the curves cross over one another (DUNLOP *et al.*, 1973; DAY, 1973; LEVI, 1974; BAILEY, 1975; JOHNSON *et al.*, 1975; LEVI and MERRILL, 1977). This is the case for many of the titanomagnetites discussed in this paper (DAY, 1973). Most of these discrepancies occur in the PSD range indicating that the transition in stability is neither abrupt nor simple, although JOHNSON *et al.* (1975) point out that crossovers could result if both fine and coarse fractions were present. SCHMIDT (1976) cast further doubt on the applicability of the test by showing theoretically that SD TRM theory predicts 'multi-domainlike' behaviour, while wall motion (MD) models predict 'single-domainlike' behaviour. Schmidt speculates on several possible explanations of this apparent paradox.

The inducing fields used for the Lowrie-Fuller test are generally about 1 Oe or greater. LEVI (1974) extended the range of inducing fields below 1 Oe and found that a critical TRM inducing field, h_c, exists ($0.10\,\mathrm{Oe} < h_c < 0.49\,\mathrm{Oe}$) such that above h_c SD behaviour is exhibited. This has implications for the Lowrie-Fuller test, but more important Levi points out that 'this suggests that the critical size for transitions of domain structure depend on the intensity of the inducing field.' The presence of an external field, therefore, can increase the effective size region of SD-PSD remanence.

The original Lowrie-Fuller test was based on the stability characteristics of TRM. However, most of the subsequent work has used ARM as a substitute for TRM. The validity of this substitution was not initially justified, but recent detailed analyses (LEVI, 1974; JOHNSON *et al.*, 1975; LEVI and MERRILL, 1976) have shown that ARM can be used to model the stability properties of TRM. However, Levi and Merrill showed that ARM was not a reasonable substitute for TRM in paleo-intensity work (for further details see LEVI and MERRILL, 1976).

5.4 *Thermal demagnetization and blocking temperatures*

The stability of TRM to thermal demagnetization is greater than that of saturation IRM (ROQUET, 1954). The saturation IRM demagnetization curve decreases

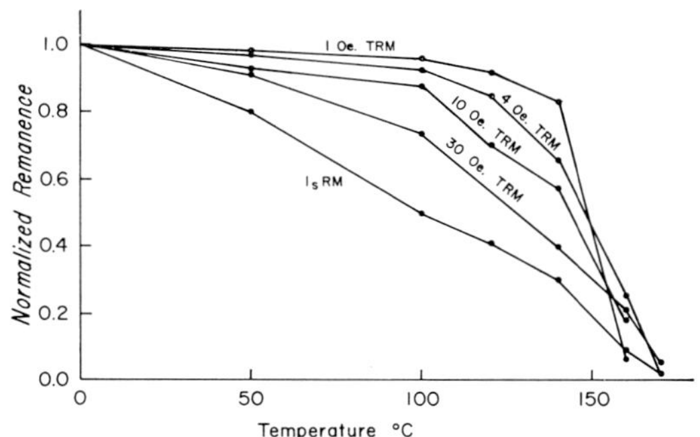

Fig. 10.   Thermal demagnetization of saturation IRM and TRMs in a titanomagnetite (composition $X=0.6$. Grain size 6.4 μm).

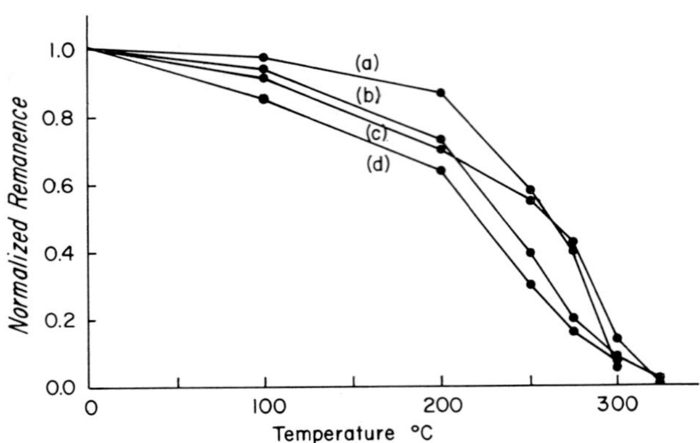

Fig. 11.   Thermal demagnetization of a 4 Oe TRM in sized titanomagnetites ($X=0.4$).

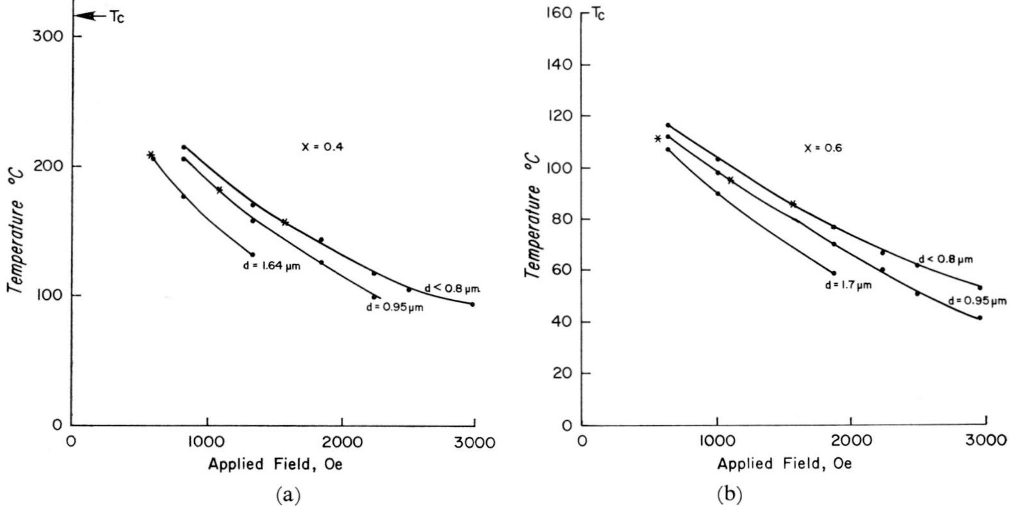

(a)   (b)

Fig. 12.   Variation of blocking temperature with applied field as deduced by Day (1975) from
anomalous thermomagnetic curves.   (a) $X=0.4$, (b) $X=0.6$.

steadily with increasing temperature showing that 'unblocking' occurs over the
whole temperature range.   TRM demagnetization curves show a progressive nar-
rowing of the blocking range as the inducing field decreases, and there is evidence
that the blocking temperatures increase with decreasing field (Roquet, 1954; and
Fig. 10).   Figure 11 shows some indication of an increase in blocking temperature
(of weak field TRM) with decreasing particle size.   This is the opposite trend to that
observed by Dunlop (1973a) in submicroscopic magnetite.

   Further evidence for the trends shown in Figs. 10 and 11 has been given by
Day (1975).   Anomalous peaks in non-saturation thermomagnetic curves were interpreted as a manifestation of the unblocking process in fine titanium rich titanomagnetites.   If this interpretation is correct then it follows (Fig. 12) that the blocking

temperature increases with both decreasing particle size and decreasing field. These measurements, however, were made in high magnetic fields. It has yet to be shown that the same phenomenon occurs in fields less than 600 Oe.

6. Evaluation of TRM Models

Theoretical models of TRM give a good qualitative explanation of the observed properties of TRM. In particular, they provide a physical basis for the blocking processes, and the high intensity and stability of TRM in SD, PSD and MD assemblages. Quantitatively, however, the models are not so successful, especially in their predictions of the induction curve. Let us now evaluate the models in the light of our recent experimental observations.

6.1 Direction and polarity

The directional property of TRM was established very early in the history of rock magnetism. More recently, DOELL and COX (1963) showed that the direction of the NRM (TRM) in recent Hawaiian lavas agreed with the geomagnetic records at the Honolulu Geomagnetic Observatory. Both theory and experiment agree that, except in the case of strongly anisotropic rocks, TRM is parallel to the inducing field.

The question of self-reversal of TRM versus field reversal was debated for many years. Although the problem has been resolved in favour of field reversals we do have several examples of self-reversals (e.g., NAGATA *et al.*, 1952; SCHULT, 1968; KROPÁČEK, 1968; WESTCOTT-LEWIS and PARRY, 1971). However, these occurrences are rare and generally confined to the hematite-ilmenite series. We should not expect our simple TRM models to include an explanation of this complicated phenomenon. Self-reversal mechanisms, however, do exist (see STACEY and BANERJEE, 1974, chapter 12).

6.2 Intensity and stability

TRM is more intense than IRM and induced magnetization (produced in the same field intensity) and much more stable than IRM. The Koenigsberger ratio, Q_t, is generally greater than 1 and increases for decreasing particle size. Incidentally, STACEY (1967) derived a theoretical value of 1/2 for Q_t in MD grains, and suggested that Q_t values greater than this indicate a contribution to the TRM from p.s.d. grains. The Q_t values in Table 6 would suggest a limiting value close to 1 for MD grains.

These properties are predicted by all TRM models, and are a direct consequence of the decrease in coercivity and the increase in thermal activation energy at elevated temperatures. IRM can only be acquired in grains whose coercive force is less than the inducing field (Section 4). Hence, only the low coercivity fraction contributes to a weak field IRM. In TRM, however, the remanence is acquired at a temperature where the coercive force has been drastically reduced so that it is now possible to activate both the low and high coercivity fractions (see, for example, SCHMIDT, 1976).

A linear increase of weak field TRM with increasing H is a property of all TRM models. For SD and PSD grains the tanh H dependence is linear in the low field limit ($\tanh X \to X$ for $X \to 0$), while for MD grains the low field theory predicts the linear dependence directly. It is very discouraging to find experimentally that weak field TRM is not linear with H. In Table 6, a linear dependence would be indicated by $n_1 = 1$. In fact, n_1 is always less than 1 and in some cases the non-linearity is severe (e.g., submicroscopic magnetite). Some possible causes of this non-linearity are given in Section 6.3.

Specific TRM (TRM/Oe) varies as d^{-l} in the PSD range where l is about 0.7 or so. Theory suggests $l = 1$ (STACEY and BANERJEE, 1974). This suggests that PSD TRM is not as simple as our present theories imply. BANERJEE (1977, this issue) discusses some of these complexities, and indicates the direction in which we should proceed to obtain a better understanding of PSD magnetization. The size dependence of weak field TRM is not too unlike that of the coercive force. Maybe we are seeing the influence of the coercive force in the TRM data, although it is not clear how the coercive force would appear in the TRM expressions.

For weak field MD TRM, theory predicts that T_B is independent of H, while for stronger inducing fields, T_B should decrease with increasing H and reach room temperature in the high field limit. This is because the TRM is now a saturation IRM. The predicted high field variation was found by ROQUET (1954) and DAY (1975, and Fig. 12). But there is some evidence (Fig. 10) that T_B varies in a similar manner in the weak field region.

It is not appropriate at the present time to look in detail at the effects of particle size and coercive force on T_B. We lack the data required to make any meaningful comparison between theory and experiment. More data is needed on the effects of H_c and size on T_B. It is surprising to find that so little information of this type (except for the early work of EVERITT, 1961, 1962a, b) has been obtained. These variations could provide some important clues to the mechanisms of TRM.

6.3   TRM induction

The TRM induction curves (Fig. 6) were obtained from SD, PSD and MD samples. However, the curves do not reflect this distinction, and MD theory provides us with the best fit to the data.

The $\tanh H$ dependence of the SD and PSD models cannot be made to fit the data. In all cases, the theoretical curve is much too steep and saturates at very low fields. Figure 13 illustrates these facts in the case of a basalt containing SD magnetite. In an attempt to reconcile this discrepancy DUNLOP and WEST (1969) extended Néel's SD theory to include the effects of interactions. This provides a much better fit (as is shown in Fig. 13). However, we have no independent evidence that these interactions exist, or if they are equally capable of improving the agreement for titanomagnetite samples where interactions are less likely to be important.

We can write Néel's SD induction equation in the form:

$$J_{\mathrm{TRM}}/J_{\mathrm{RS}} = \tanh \frac{V J_s(T_B) H}{k T_B}$$

where J_{RS} is the saturation IRM.

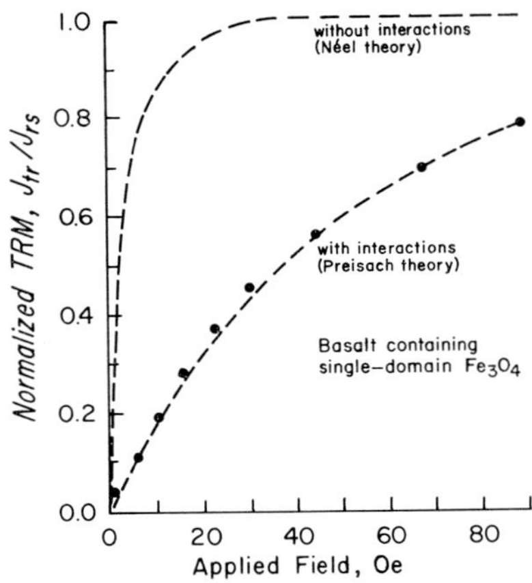

Fig. 13.   Influence of particle interactions on the intensity of TRM (after DUNLOP and WEST, 1969).

From this equation we see that it is the parameters in the argument of the tanh function that must be investigated if we wish to improve the agreement.   Dunlop and West modified H by introducing an interaction field.   We should also investigate the exact nature of $J_s(T_B)$ and T_B.   Again, we see that we need more T_B data.   Notice that the volume, V, is present in the tanh function.   If we overestimate the grain size, the error is magnified when calculating the volume.   In weak fields, the induction equation reduces to

$$J_{\mathrm{TRM}}/J_{\mathrm{RS}} \simeq \frac{V J_s(T_B)}{k T_B} H$$

showing that the observed non-linearity of weak field TRM could also be due to variations in these parameters (including of course the introduction of an interaction field).

The best fit to the TRM induction data is provided by the MD models, but even then there are some disturbing features:

1)   The non-linearity in the low field region ($n_1 < 1$) still cannot be explained.

2)   The transition field, H_{12}, between low field and high field data does not depend on H_c (DAY, 1973) or H_f (DUNLOP and BINA, 1977).   Theory predicts a strong dependence on H_c and H_f.

3)   There is no explanation for the third linear segment that occurs in the high field region.

4)   The MD models are equally successful in fitting the SD and PSD TRM data. DUNLOP et al. (1974) have attempted to improve the fit in the weak field region by introducing a PSD contribution.   PSD TRM saturates in low fields (see Fig. 8 in DUNLOP, 1977; this issue) so that a TRM induction equation:

$$J_{\mathrm{TRM}} = AH + BF(\alpha H)$$

could improve the fit for weak field TRM.   In this equation the first term represents the MD contribution and the second term the PSD contribution. $F(\alpha H)$ is a function derived by Stacey and Banerjee for PSD TRM (a table of $F(a)$ versus a is given by STACEY and BANERJEE, 1974). However, in spite of having two curve fit parameters (A/B and α), the fit is not significantly better.

MD theory allows a greater flexibility in the value of the intermediate slope, n_2. In the MD models, n_2 is a measure of the power index in the relation

$$H_c = \beta J_s^n \qquad n_2 = 1 - \frac{1}{n}.$$

Any value of n greater than 1 is permissible and n can conceivably vary over a large range ($1 < n < 10$) depending on the origin of the coercivity. The n values calculated from the induction curves (n_{calc} in Table 6) are in the range of $1.28 < n_{\text{calc}} < 2.94$. Dunlop (see DUNLOP, 1973a; and DUNLOP and WADDINGTON, 1975) has obtained n directly from measurements of H_c at elevated temperatures (Table 6). There is a reasonable agreement between n and n_{calc}, but notice that the synthetic magnetite give values of n that are higher than n_{calc} while the opposite is found for the rock samples.

The transition field, H_{12}, shows no significant dependence on H_c. This is very troublesome because Schmidt's theoretical induction curves (Fig. 2 in SCHMIDT, 1976) show a marked dependence. This problem appears to be insurmountable within the framework of present MD models.

There is no explanation in the MD model for the high field behaviour. It might reflect changes in the domain configuration brought about by the large inducing fields, i.e., it could be associated with some approach to saturation process. We have not included Merrill's modifications of the demagnetizing field in the MD model. The transition field H_{23} may be related to the change in the form of the demagnetizing field. However, if this were the case we would have to use a modified induction curve in the intermediate region. This would destroy the agreement between n and n_{calc}. Also, one would intuitively expect this changeover to occur at much lower fields.

Last but not least, there is the disturbing observation that SD and MD TRMs behave in a very similar manner. TRM induction curves therefore give no clues as to how to distinguish between the TRM in SD, PSD and MD particles. As I see it, we have two choices. First, we have to consider the possibility that bilogarithmic plots of TRM data destroy any features that may be present. There are a number of reasons for suggesting this:

1)　The transition of H_{12} is too sharp. One would have expected that variations in grain size and coercive force would preclude the possibility of a sharp transition.

2)　Bilogarithmic plots can easily misrepresent experimental data. For example, the data points in Fig. 14 would be equally spaced on a linear plot.

3)　The bilogarithmic plot of the tanh function (Fig. 14) is very similar to the observed induction curves. It is not hard to persuade oneself that there are three linear segments in Fig. 14. Maybe our data is not precise enough to distinguish be-

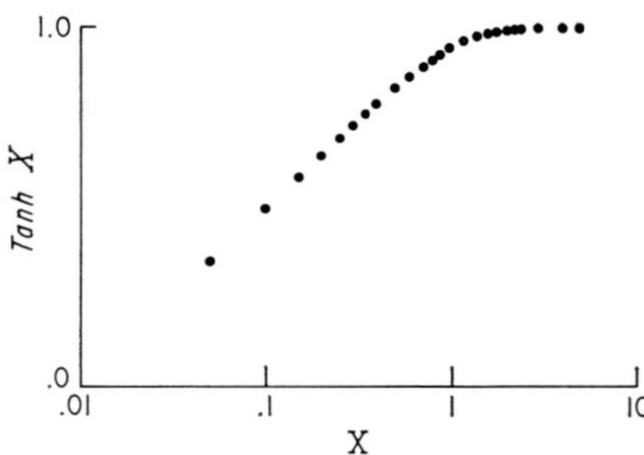

Fig. 14.   Bilogarithmic plot of the tanh function (see text).

tween a tanh H dependence and a power law dependence. Alternatively, the tanh H dependence could be intrinsically correct, in which case the problem lies in the argument of the tanh function.

4)   It is difficult to accept the fact that TRM acquisition in SD, PSD and MD samples is so similar.

The other alternative is to accept the fact that the TRMs are similar and look for possible explanations. DUNLOP and WADDINGTON (1975) have suggested that strong particle interactions in SD samples can simulate the effect of an internal demagnetizing field. This could produce a 'multi-domainlike' induction curve.

Maybe we have underestimated the MD model. It could apply far beyond the limits of the two domain particles on which it is nominally based. There are no experimental observations that give any of the SD-MD transition with temperature. Simple minded calculations (DAY, 1973), based on the methods discussed in Section 2.1, predict a decrease in the transition size with increasing temperature so that a grain could conceivably be SD at room temperature but MD at T_B. If this high temperature configuration became blocked then we could produce a MD induction curve. These calculations, however, ignore the important fact that the domain wall width increases with temperature also. With BUTLER and BANERJEE's (1975) refinements we find that the opposite is true, i.e., the SD-MD transition increases with increasing temperature. In either case, it could affect the TRM mechanism, and the domain wall expansion at elevated temperatures should certainly be investigated. These points are discussed further by DAY (1973), LEVI (1974), BUTLER and BANERJEE (1975), and LEVI and MERRILL (1976).

The third possibility is an as-yet-undiscovered TRM mechanism that is capable of producing the observed universal TRM induction curve!

7.   Conclusion

We have seen that although we can explain many of the observed properties of TRM we still do not have a complete theory. It is very discouraging to find that our models are the least successful in the weak field region. This is the region which is most important for paleomagnetism studies.

The problems that we have encountered in our evaluation of TRM should now be used to point the way for future research. The most important experiments to be done involve measurements at elevated temperatures. Knowledge of the variations of domain structure, domain state and hysteresis properties with temperature is extremely important for a deeper understanding of TRM. The effect of coercive force and grain size on T_B will provide clues to the blocking mechanism.

These observations are vital in our search for a complete theory of TRM.

I would like to thank the following people for their help and encouragement over the past several years: Hal Noltimier, Bill O'Reilly, Walter Pilant and Vic Schmidt. This paper was improved significantly by the detailed reviews of Shaul Levi, Dave Dunlop and Ron Merrill, the patience of Diane Mondragon and Elizabeth Mendly, and the competent drafting of Dave Crouch. Finally I would like to express my gratitude to Mike Fuller, whose help and advice over the years could never be measured. This work was funded by grants from the National Science Foundation.

REFERENCES

AMAR, H., On the width and energy of domain walls in small multi-domain particles, *J. Appl. Phys.*, **28**, 732–733, 1957.

AMAR, H., Size dependence of the wall characteristics in a two-domain iron particle, *J. Appl. Phys.*, **29**, 542–543, 1958a.

AMAR, H., Magnetization mechanism and domain structure of multi-domain particles, *Phys. Rev.*, **111**, 149–153, 1958b.

BAILEY, M.E., The magnetic properties of PSD grains, M.S. thesis, University of Toronto, 1975.

BAILEY, M.E., D.J. DUNLOP, and K.L. BUCHAN, Alternating field demagnetization characteristics of pseudo-single-domain grains, *EOS*, **56**, 353, 1975.

BANERJEE, S.K., On the origin of stable remanence in pseudo single-domain grains, *J. Geomag. Geoelectr.*, **29**, 319–329, 1977.

BEAN, C.P. and J.D. LIVINGSTON, Superparamagnetism, *J. Appl. Phys.*, **30**, 120S–129S, 1959.

BERKOWITZ, A.E. and W.J. SCHUELE, Magnetic properties of some ferrite micropowders, *J. Appl. Phys.*, **30**, 134S, 1959.

BROWN, W.F., Relaxational behavior of fine magnetic particles, *J. Appl. Phys.*, **30**, 130S, 1959.

BUTLER, R.F. and S.K. BANERJEE, Theoretical single-domain grain size range in magnetite and titanomagnetite, *J. Geophys. Res.*, **80**, 4049–4058, 1975.

DAY, R., The effect of grain size on the magnetic properties of the magnetite-ulvospinel solid solution series, Thesis, Univ. of Pittsburgh, Pittsburgh, Pa., 1973.

DAY, R., Some curious thermomagnetic curves and their interpretation, *Earth Planet. Sci. Lett.*, **27**, 95, 1975.

DAY, R., M.D. FULLER, and V.A. SCHMIDT, Magnetic hysteresis properties of synthetic titanomagnetites, *J. Geophys. Res.*, **81**, 873, 1976.

DAY, R., M.D. FULLER, and V.A. SCHMIDT, Hysteresis properties of titanomagnetites: Grain size and composition dependence, *Phys. Earth Planet. Inter.*, **13**, 260, 1977.

DICKSON, G.O., C.W.F. EVERITT, L.G. PARRY, and F.D. STACEY, Origin of thermoremanent magnetization, *Earth Planet. Sci. Lett.*, **1**, 222, 1966.

DOELL, D.D. and A. COX, The accuracy of the Paleomagnetic method as evaluated from historic Hawaiian lava flows, *J. Geophys. Res.*, **68**, 1997, 1963.

DUNLOP, D.J., Hysteresis properties of synthetic and natural mono-domain grains. *Philos. Mag.*, **19**, 329, 1969.

DUNLOP, D.J., Thermoremanent magnetization in submicroscopic magnetite, *J. Geophys. Res.*, **78**, 7602–7613, 1973a.

DUNLOP, D.J., Superparamagnetic and single-domain threshold sizes in magnetite, *J. Geophys. Res.*, **78**, 1780, 1973b.

DUNLOP, D.J., Thermal fluctuation analysis; a new technique in rock magnetism, *J. Geophys. Res.*, **81**, 3511, 1976.

DUNLOP, D.J., The hunting of the 'Psark', *J. Geomag. Geoelectr.*, **29**, 293–318, 1977.

DUNLOP, D.J. and M.-M. BINA, The coercive force spectrum of magnetite at high temperatures: evidence for thermal activation below the blocking temperature, *Geophys. J.R. Astron. Soc.*, 1977 (in press).

DUNLOP, D.J. and E.D. WADDINGTON, The field dependence of thermoremanent magnetization of igneous rocks, *Earth Planet. Sci. Lett.*, **25**, 11, 1975.

DUNLOP, D.J. and G.F. WEST, An experimental evaluation of single-domain theories, *Rev. Geophys.*, 7, 709, 1969.

DUNLOP, D.J., J.A. HANES, and L.K. BUCHAN, Indices of multi-domain magnetic behavior in basic igneous rocks: alternating field demagnetization, hysteresis, and oxide petrology, *J. Geophys. Res.*, **78**, 1387, 1973.

DUNLOP, D.J., F.D. STACEY, and D.E.W. GILLINGHAM, The origin of thermoremanent magnetization: contribution of pseudo single-domain magnetic moments, *Earth Planet. Sci. Lett.*, **21**, 288, 1974.

EVANS, M.E., Single-domain particles and TRM in rocks, *Comments Earth Sci. Geophys.*, **2**, 139, 1972.

EVANS, M.E. and M.L. WAYMAN, An investigation of small magnetic particles by means of electron microscopy, *Earth Planet. Sci. Lett.*, **9**, 365–370, 1970.

EVERITT, C.W.F., Thermoremanent magnetization, I: Experiments on single-domain grains, *Philos. Mag.*, **6**, 713, 1961.

EVERITT, C.W.F., Thermoremanent magnetization, II: Experiments on multi-domain grains, *Philos. Mag.*, **7**, 583, 1962a.

EVERITT, C.W.F., Thermoremanent magnetization, III: Theory of multi-domain grains, *Philos. Mag.*, **7**, 599, 1962b.

FREI, E.H., S. SHTRIKMAN, and D. TREVES, Critical size and nucleation field of ideal ferromagnetic particles, *Phys. Rev.*, **106**, 446–455, 1957.

JOHNSON, H.P., W. LOWRIE, and D.V. KENT, Stability of anhysteretic remanent magnetization in fine and coarse magnetite and maghemite particles, *Geophys. J.R. Astron. Soc.*, **41**, 1, 1975.

KITTEL, C., Physical theory of ferromagnetic domains, *Rev. Mod. Phys.*, **21**, 541–583, 1949.

KNELLER, E.F. and F.E. LUBORSKY, Particle size dependence of coercivity and remanence of single-domain particles, *J. Appl. Phys.*, **34**, 656, 1963.

KOBAYASHI, K. and M.D. FULLFR, Stable remanence and memory of multi-domain materials with special reference to magnetite, *Philos. Mag.*, **18**, 601, 1968.

KOENIGSBERGER, J.G., Natural residual magnetism of eruptive rocks, *Terr. Magn. Atmos. Electr.*, **43**, 119 and 299, 1938.

KROPÁČEK, V., Self-reversal of spontaneous magnetization of natural cassiterite, *Stud. Geophys. Geod. Českoslov. Acad. Ved.*, **19**, 108, 1968.

LARSON, E., M. OZIMA, M. OZIMA, T. NAGATA, and D. STRANGWAY, Stability of remanent magnetization of igneous rocks, *Geophys. J.R. Astron. Soc.*, **17**, 263–292, 1969.

LEVI, S., Some magnetic properties of magnetites as a function of grain size and their implications for paleomagnetism, Ph.D. thesis, Univ. of Washington, 1974.

LEVI, S. and R.T. MERRILL., A comparison of ARM and TRM in magnetite, 1976 (in press).

LEVI, S. and R.T. MERRILL, Properties of single-domain, pseudo single-domain and multi-domain magnetite, *J. Geophys. Res.*, 1977 (in press).

LOWRIE, W., The effects of internal stress on remanence and coercive force in nickel and magnetite, Thesis, University of Pittsburgh, Pittsburgh, Pa., 1967.

LOWRIE, W. and M.D. FULLER, On the alternating field demagnetization characteristic of multi-domain magnetization in magnetite, *J. Geophys. Res.*, **76**, 6339, 1971.

LUBORSKY, F.E., Development of elongated fine particle magnets, *J. Appl. Phys.*, **32**, 1715, 1901.

MEIKELJOHN, W.H., Experimental study of the coercive force of fine particles, *Rev. Mod. Phys.*, **25**, 302, 1953.

MELLONI, M., *Napali, Atti. Acc. Sci.*, **1**, 121, 1853.

MERRILL, R.T., The demagnetization field of multi-domain magnetite, *J. Geomag. Geoelectr.*, **29**, 285–292, 1977.

MORRISH, A.H. and S.P. YU, Dependence of the coercive force on the density of some iron oxide powders, *J. Appl. Phys.*, **26**, 1049–1055, 1955.

MURTHY, G.S., M.E. EVANS, and D.I. GOUGH, Evidence for single-domain magnetite in the Michikamau anorthosite, *Can. J. Earth Sci.*, **8**, 361–370, 1971.

NAGATA, T., The mode of causation of thermo-remanent magnetism in igneous rocks, Preliminary note, *Bull. Earthq. Res. Inst.*, **19**, 49, 1941, and **20**, 192, 1942.

NAGATA, T., S. UYEDA, and S. AKIMOTO, Self-reversal of thermoremanent magnetization of igneous rocks, *J. Geomag. Geoelectr.*, **4**, 22, 1952.

NÉEL, L., Propriétés d'un ferromagnétique cubique en grains fins, *C.R. Acad. Sci.*, **224**, 1498, 1947.

NÉEL, L., Théorie du traînage magnétique des ferromagnétiques en grains fins avec applications aux terres cuites, *Ann. Géophys.*, **5**, 99, 1949.

NÉEL, L., Some theoretical aspects of rock magnetism, *Adv. Phys.*, **4**, 191, 1955.

OZIMA, M. and M. OZIMA, Origin of thermoremanent magnetization, *J. Geophys. Res.*, **70**, 1363, 1965.

OZIMA, M., M. OZIMA, and T. NAGATA, Low-temperature treatment as an effective means of "magnetic cleaning" of natural remanent magnetization, *J. Geomag. Geoelectr.*, **16**, 37, 1964.

PARRY, L.G., Magnetic properties of dispersed magnetite powders, *Philos. Mag.*, **11**, 303–311, 1965.

PETROV, I.N. and V.V. METALLOVA, 12V., *Earth Sci.*, **9**, 555, 1968.

RAHMAN, A.A., A.D. DUNCAN, and L.G. PARRY, Magnetisation of multi-domain magnetite, *Riv. Ital. Geofis.*, **22**, 259, 1973.

RIMBERT, F., Contribution a l'etude de l'action de champs alternatifs sur les aimanations remanentes des roches. Applications geophysiques, *Rev. Inst. Fr. Pét.*, **14**, 17 and 123, 1959.

ROBINS, B.W., Remanent magnetization in spinel iron-oxides, Ph.D. thesis, University of New South Wales, Australia, 1972.

ROQUET, J., Sur les rémanances magnétiques des oxydes de fer et leur intérêt en géomagnétisme, *Ann. Géophys.*, **10**, 226 and 182, 1954.

SCHMIDT, V.A., A malti-domain model of thermoremanence, *Earth Planet. Sci. Lett.*, **20**, 440–446, 1973.

SCHMIDT, V.A., The variation of the blocking temperature in models of thermoremanence (TRM), *Earth Planet, Sci. Lett.*, **29**, 146, 1976.

SCHULT, A., Self-reversal of magnetization and chemical composition of titanomagnetites in basalts, *Earth Planet. Lett.*, **4**, 440, 1968.

SHIVE, P.N., Dislocation control of magnetization, *J. Geomag. Geoelectr.*, **21**, 519, 1969a.

SHIVE, P.N., The effect of internal stress on the thermoremanence of nickel, *J. Geophys. Res.*, **74**, 381, 1969b.

SOFFEL, H., The single-domain/multi-domain transition in intermediate titanomagnetites, *Z. Geophys.*, **37**, 451, 1971.

SOFFEL, H., Domain structure of titanomagnetites and its variation with temperature, *J. Geomag. Geoelectr.*, **29**, 277–284, 1977.

STACEY, F.D., Thermoremanent magnetization (TRM) of multi-domain grains in igneous rocks, *Philos. Mag.*, **3**, 1391, 1958.

STACEY, F.D., A generalized theory of thermoremanence, covering the transition from single-domain to multi-domain magnetic grains., *Philos. Mag.*, **7**, 1887, 1962.

STACEY, F.D. and S.K. BANERJEE, *The Physical Principles of Rock Magnetism*, pp. 195, Elsevier, New York, 1974.

STACEY, F.D. and K.N. WISE, Crystal dislocations and coercivity in fine-grained magnetite, *Aust. J. Phys.*, **20**, 507, 1967.

STONER, E.C. and E.P. WOHLFARTH, A mechanism of magnetic hysteresis in heterogeneous alloys. *Philos. Trans. R. Soc. London, Ser. A*, **240**, 599–642, 1948.

SYONO, Y., S. AKIMOTO and T. NAGATA, Remanent magnetization of ferromagnetic single crystal, *J. Geomag. Geoelectr.*, **14**, 113, 1962.

SYONO, Y., Magnetocrystalline anisotropy and magnetization of Fe_3O_4-Fe_2TiO_4 series with special application to rock magnetism, *Jpn. J. Geophys.*, **4**, 71, 1965.

THELLIER, E., Sur l'Aimantations des ferres cuites et ses applications geophysiques, *Ann. Inst. Phys. Globe, Univ. Paris*, **16**, 157, 1938.

UYEDA, S., TRM as a medium of paleomagnetism, with special reference to reverse TRM, *J. Geomag. Geoelectr.*, **2**, 1, 1958.

VERHOOGEN, J., The origin of thermoremanent magnetization, *J. Geophys. Res.*, **64**, 2441, 1959.

WASILEWSKI, P.J., Magnetic hysteresis of natural materials, *Earth Planet. Sci. Lett.*, **20**, 67, 1973.

WESCOTT-LEWIS, M.F. and L.G. PARRY, Thermoremanence in synthetic rhombohedral iron-titanium oxides, *Aust. J. Phys.*, **24**, 735, 1971.

Adv. Earth Planet. Sci., **1**, 35–43, 1977

Single Domain Oxide Particles as a Source of Thermoremanent Magnetization

M.E. EVANS

Institute of Earth and Planetary Physics, University of Alberta,
Edmonton, Alberta, Canada

(Received June 2, 1977)

Single domain (SD) particles have long been suggested as a potentially strong and stable source for the paleomagnetic signal, but their actual occurrence in rocks has been much questioned. Two possible modes of occurrence of SD Fe-Ti oxides can be distinguished. These are as magnetite rods bounded by ilmenite lamellae in intergrown grains (analogous to Alnico permanent magnet alloys), and as isolated ultrafine particles, perhaps representing part of a much broader grain-size distribution (analogous to the oxide coatings used in tape recording). The evidence for the presence in rocks of these two types of magnetic particle is reviewed. Where intergrown grains are present the SD 'Alnico model' seems to be valid, and where isolated titanomagnetite grains are involved the SD 'tape-recorder model' is satisfactory. Where pure Fe_3O_4 particles are involved the situation is less clear, and current thinking stresses the role of the so-called pseudo-single domain (PSD) moments.

1. Introduction

The search for an adequate explanation of the origin of the paleomagnetic signal has had a long and interesting history. In terms of igneous rocks, which have played a major role in paleomagnetic research, there is no doubt that some form of thermoremanence (TRM) is dominant. NÉEL (1949) originally proposed an elegant single domain (SD) theory, but later (NÉEL, 1955) came to favour multidomain (MD) particles as the dominant TRM carriers on the grounds that the average particle size encountered in rocks is much too large to permit SD behaviour. Nevertheless, the TRM's carried by most rocks persistently exhibit SD-like properties—in particular high coercivities (DUNLOP, 1973). Furthermore, pure MD theories proved inadequate (STACEY and BANERJEE, 1974, p. 110), and the pendulum returned towards SD theory, aided by observational evidence that sufficiently small magnetic particles may, in fact, occur in rocks. Currently, there is much support for a compromise situation which stresses the role of particles lying in the transition region between true SD and true MD behaviour. These so-called pseudo-single domain (PSD) effects were first invoked by STACEY (1961), and are reviewed by BANERJEE (1977) and by DUNLOP (1977).

Two problems have persistently hampered progress. Firstly, it has proved difficult to establish a complete understanding of the physics of PSD particles, to the extent that an adequate theory it still lacking (DUNLOP, 1977). Secondly, the critical particle size for true SD behaviour in magnetite is less than the wavelength of light ($\sim0.5\ \mu$m), which renders SD particles very difficult to locate and identify. This paper reviews and discusses the data pertinent to the second of these problems.

2.   Naturally Occurring Ultrafine Particles

Despite Néel's (1955) original doubts, it now appears that particles small enough to be single domains do, in fact, occur in rocks, although only a few cases have been studied in detail. This situation contrasts with that found in technological applications such as tape recording and permanent magnet materials, where a vast literature now exists (e.g., Bate, 1975; Hadfield, 1962). In terms of the materials found in nature it is convenient to distinguish two modes of occurrence for SD particles; individual ultrafine particles, perhaps representing one end of a continuous distribution, and as small regions of larger grains which are subdivided by intergrowth structures of two or more phases. These two modes are discussed separately.

2.1   Intergrowths

Magnetite and ilmenite commonly occur together in large grains (10–100 μm) intimately intergrown on a submicron scale. The ilmenite forms an octahedral pattern of lamellae developed on the (111) planes of the magnetite. Such grains have an appearance not unlike the so-called Widmanstätten patterns observed in iron meteorites. This form of subdivision was first put forward as a source of high coercive forces, and thus of stable remanence, by Graham in 1953, and was further supported by Powell (1963) and Larson et al. (1969). It has a close analogue in the development of permanent magnet materials, since Alnico alloys (which account for 50% of the U.S. permanent magnet market, Cullity, 1972), are essentially two-phase materials consisting of a series of magnetic rods embedded in a non-magnetic matrix. The desirable properties of these materials are attributed to the microstructure revealed by electron microscopy (de Vos, 1969) and in particular, the magnetic hardness is caused by 'the shape anisotropy of singledomain particles' (Cullity, 1972, p. 568).

In samples of geophysical interest the control of coercivity by subdivision is now well established (see Table 1 for summary), and the techniques necessary to reveal microstructure are available (e.g., see Hoblitt and Larson, 1975). However, the corresponding domain state has proved more difficult to firmly establish. Strangway et al. (1968) attributed the very high stability found in many volcanic rocks to the presence of elongated volumes of magnetite bounded by ilmenite lamellae. They suggested that some of these will be single domains whose shape anisotropy gives rise to high coercivities. The actual model they used to calculate critical SD sizes is unphysical in that the domain walls they envisaged were not true Bloch walls. Nevertheless an improvement suggested by Murthy et al. (1971) indicated that the actual critical sizes obtained were not much in error. Stacey and Banerjee (1974) considered sheet-like lamellar intergrowths and concluded that 'for all macroscopic prop-

Table 1.   Various indicators of magnetic hardness in homogeneous and intergrown grains.*

Authors	'Hardness' indicator	Sample	Homogeneous	Intergrown
STRANGWAY *et al.* (1968)	% of 1.2 Oe TRM remaining after 1,000 Oe AF demagnetization	WR-3-4-2 WR-1-2-2 PJ-16-3-3 PJ-11-3-3	4.2% 7.2% 8.9% 9.3%	12.7% 22.6% 13.1% 14.5%
LARSON *et al.* (1969)	Coercive force (H_c)	sample 13 (basalt)	45 Oe	180 Oe
	Remanent coercivity (H_{CR})	"	108 Oe	386 Oe
EVANS and WAYMAN (1974)	MDF of 1.2 Oe ARM	magnetite-ulvospinel	37 Oe	170 Oe
DAVIS and EVANS (1976)	MDF of 1.2 Oe ARM	magnetite-ilmenite	60 Oe	315 Oe
MANSON and O'REILLY (1976)	Coercivity spectra peaks derived from IRM induction	sample RK (basalt)	$\sim$200 Oe	700 Oe

 *   This table is intended simply to illustrate the relationship between microstructure and magnetic hardness, and is by no means exhaustive. For fuller discussions and other examples reference should be made to the original papers.

erties we may treat them as multidomains.' This may be true for strictly sheet-like intergrowths, but such regions will only occur if ilmenite development is restricted to one of the four possible (111) planes. This is commonly not the case, although a detailed study is required to firmly establish the geometry actually realized as crystallization and/or oxidation proceeds. DAVIS and EVANS (1976) investigated intergrown grains produced by in-air heating of a Tertiary basalt. They inferred that the magnetically dominant magnetite regions occur as rods. A high value of 0.3 for the ratio saturation remanence/saturation magnetization (J_{RS}/J_S) was obtained, and the reduction from the theoretical value of 0.5 for randomly oriented SD particles was found to be quantitatively explicable in terms of local demagnetizing fields. Other magnetic properties were closely similar to those of a sample containing synthetic acicular particles of SD Fe_3O_4. It was therefore concluded that, for paleomagnetic purposes, intergrown grains behave as arrays of interacting SD particles. Similar results were obtained for the magnetite/ulvospinel intergrowths studied earlier by EVANS and WAYMAN (1974). In both of these cases, the intergrowth structure was shown by electron microscopy to be developed on a scale well below 1 μm. MANSON and O'REILLY (1976) have also studied the change in magnetic properties as homogeneous titanomagnetites in basalts are progressively oxidized. They carefully monitored the resulting changes by electron microscopy and measurements of rotational hysteresis and coercivity spectra. They concluded that phase-splitting generates 'Fe-rich needles' which give rise to high coercivities and cause a dramatic increase in the height of the rotational hysteresis peak. The microstructure never became

resolvable by optical microscopy and the grains therefore remained in Class I of
WILSON and WATKINS (1967). Moreover, the magnetic changes were well in progress before any observable microstructure could be detected even by electron microscopy. This finding concurs with the results of CREER and PETERSEN (1969) who
appealed to submicroscopic exsolution to explain the magnetic properties of oxidized
basalts. One is also reminded of the conclusions reached by RADHAKRISHNAMURTHY
and DEUTSCH (1974) who invoke 'some yet unknown mechanism of subdivision' to
explain magnetic evidence for the presence of superparamagnetic (SP) particles in
basalts. In some cases, magnetic interaction within dense clusters of SP particles
causes finite values of coercive force and remanence to reappear (RADHAKRISHNAMUR-
THY et al., 1973; EVDOKIMOV, 1963), thus enabling such material to carry paleomagnetic information.

Another class of intergrowths which have been investigated in some detail are
the Widmanstätten patterns found in iron meteorites. Interest in these extraterrestrial
objects stems from the possibility that they carry a record of interplanetary magnetic
fields which may have played an important role in the early development of the solar
system. Compared to terrestrial basalts the Widmanstätten intergrowths are very
coarse (~ 1 mm) and the magnetism correspondingly soft. However, some areas are
subdivided on a finer scale, and the micro-kamacite regions in these so-called plessite areas appear to be singledomain grains capable of carrying a stable remanence
(BRECHER and CUTRERA, 1976). It is important to realize that although microkamacite regions appear capable of carrying a stable remanence it is apparently still
an open question as to whether they actually do retain a memory of their early
magnetic environment (BRECHER and ALBRIGHT, 1977).

2.2 Distribution of isolated particles

STRANGWAY (1961) suggested that the source of stable remanence in certain
Canadian diabases might be what he termed a 'powder distribution' of very fine
opaque particles occurring in pyroxene and olivine grains. Subsequently EVANS and
MCELHINNY (1969) and EVANS et al. (1968) suggested that at least some of the stable
remanence of the very ancient (2600 my) Modipe Gabbro of southern Africa resides
in minute, often elongated, magnetite particles exsolved within pyroxene grains.
Similar suggestions involving opaque grains in feldspar crystals have been put forward by HARGRAVES and YOUNG (1969) and MURTHY et al. (1971).

Since the resolving power of optical microscopes is inadequate to penetrate the
'SD barrier' further work has inevitably required the application of electron microscopy. EVANS and WAYMAN (1970) followed up the work on the Modipe Gabbro
by applying a two-stage replication technique in an attempt to discover if the particle size distribution continues below the optical limit. It was demonstrated that
optically unresolvable particles are plentiful and the authors concluded that SD
magnetite particles can, and do, occur naturally in rocks (see EVANS and WAYMAN,
1970, Fig. 6). Four words of warning are appropriate. Firstly, as in all microscopy, a two-dimensional cross section yields no information about the third dimension

of each particle. Secondly, a single demonstration case does not provide sufficient basis for a meaningful assessment of the overall significance of the phenomenon. Thirdly, no claim can be made that the minute part of the sample studied is at all representative of the entire rock. Fourthly, replication techniques can resolve ultra-fine textures but yield very little compositional information, being limited to what can be inferred from the action of the etchant employed.

Evidence for SD particles occurring in the interior of silicate grains in also provided by the work of HOYE and O'REILLY (1973) and HOYE and EVANS (1975). These investigations concern the growth of iron oxide particles (thought to be Fe_3O_4) by oxidation of synthetic olivines $(Fe, Mg)_2SiO_4$. Acquisition of chemical remanent magnetization (CRM) as the magnetic particles nucleate and grow was monitored and found to pass through a peak after several hundred minutes oxidation time. This observation parallels that found by KOBAYASHI (1961) who investigated CRM in a Cu–Co alloy and reports a similar, but much sharper, peak. In both cases the shape of the CRM vs oxidation time curve reflects the gradual growth of particles from superparamagnetic (SP) to SD and thence to PSD (or multidomain, MD) con-figurations. The occurrence of ultrafine iron oxide particles in oxidized olivines is confirmed by the work of CHAMPNESS (1970), both on mineralogical grounds and by direct observation using high voltage transmission electron microscopy.

3.   Discussion

Two modes of occurrence for ultrafine magnetic particles capable of carrying a strong and stable paleomagnetic signal can be distinguished. In terms of tech-nological developments these correspond to Alnico permanent magnets (magnetite/ilmenite intergrowths), and to oxide coatings employed in tape recording (isolated ultrafine particles of magnetite or titanomagnetite).

The 'Alnico model' is very attractive—the microstructural control of mag-netic hardness in natural samples is well established (Table 1), and there is good evidence for the presence of rod-like magnetite regions acting as SD particles. This SD model is further strengthened by the observation that the high packing factors found in intergrown grains can be expected to somewhat increase the critical size necessary for SD behaviour (MORRISH and WATT, 1957). Apparently, even when no subdivision is visible SD properties may appear (RADHAKRISHNAMURTHY and DEUTSCH, 1974; MANSON and O'REILLY, 1976). Strong support for these findings is provided by the work of JENSEN and SHIVE (1973) who analysed Mössbauer spectra of synthetic titanomagnetites and demonstrated that the titanium ions tend to cluster together and thereby introduce inhomogeneity which they suggest might be regarded as incipient exsolution.

The 'tape-recorder model' has received support from electron microscopy (EVANS and WAYMAN, 1970) and from Bitter pattern studies (SOFFEL, 1968, 1969, 1971). Single-domain magnetite and titanomagnetite particles do apparently occur naturally in at least some rock samples. They can also be generated in synthetic silicate samples by suitable laboratory treatment (HOYE and O'REILLY, 1973). When present they represent a potent source of strong, stable remanence which 'have a disproportionate effect on remanent properties' (DUNLOP and WEST, 1969). For example, STRANGWAY et al. (1968) argue that SD volume fractions as small as a few ppm suffice to account for typical NRM values of strongly magnetized rocks such as basalts. However, RYALL and ADE-HALL (1975) have pointed out that although TRM per unit volume increases dramatically as grain size decreases, when the actual volume of individual particles is taken into account each large particle contributes as much to the total TRM as do many small particles. They show that the grain size distribution of EVANS and MCELHINNY (1969) combined with the experimental TRM vs grain size curve (DUNLOP, 1973) indicates that PSD grains a few microns in size dominate the total TRM of certain oceanic basalts. There are several objections to their procedure:

1.   The applicability of a particle size distribution determined from a Precambrian gabbro to a Tertiary oceanic basalt is highly questionable.

2.   The grain counts reported by EVANS and MCELHINNY (1969) were intended simply to indicate that the size distribution continues below the optical limit. It is inadmissible to use data derived from optical microscopy to represent the quantative significance of sub-microscopic particles. Even the electron microscope results of EVANS and WAYMAN (1970) are inadequate in this respect—they simply confirm the extension of the size distribution below the limit of optical resolution. No claim that these measurements were representative was ever made or implied.

3.   The situation discussed by RYALL and ADE-HALL (1975) is much closer to that reported in a very important series of papers by SOFFEL (1968, 1969, 1971). In the present context two major conclusions emerge from Soffel's work on German Tertiary basalts. Firstly, after a thorough study of the magnetic properties and domain configurations, including direct Bitter pattern observations, SOFFEL (1971, p. 207) concludes that 'the stable component of the TRM of the two basalts is located entirely in the small particles with less than 1 micron diameter having single domain configuration.' Secondly, the actual critical size for SD behaviour in the titanomagnetites is considerably larger than for pure Fe_3O_4. SOFFEL (1971, p. 469) concludes that for titanomagnetites of composition $0.55Fe_2TiO_4$–$0.45Fe_3O_4$ 'most of the particles with diameters of 1.5 microns or less must be regarded as single domain particles.' Thus the conclusion of Ryall and Ade-Hall that optically visible grains dominate the remanence does not conflict with a SD origin of TRM. This view is strengthened by the extremely re-entrant nature of the surfaces of the typical dendritic opaques found in oceanic basalts. Morphologies of this kind imply that although a grain may have an overall size of several microns it effectively consists of several smaller units.

Despite these objections to the specific treatment given by RYALL and ADE-HALL (1975), their general point concerning the actual magnetic moments of particles merits attention. The point concerns the amount of TRM contributed by SD particles compared to that of PSD and MD grains. STRANGWAY et al. (1968) concluded

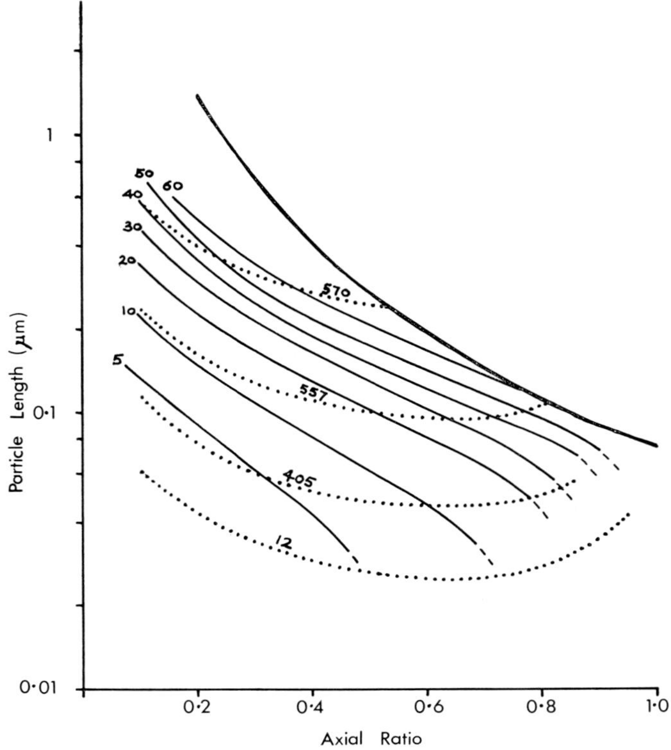

Fig. 1.   TRM of single domain magnetite grains, obtained from NÉEL's (1949) theory.   Numbers on the solid curves represent TRM in emu· cm^{-3}·Oe^{-1}.   The dotted curves and the numbers on them illustrate several blocking temperatures (°C) for relaxation times of 100 seconds. Computations assumed the dependence of spontaneous magnetization on temperature given by PAUTHENET (1952).   The solid heavy curve shows the SD threshold given by BUTLER and BANERJEE (1975).

that a volume fraction of magnetite on the order of a few ppm suffices to account for typical TRM's encountered in basalts, and EVANS and MCELHINNY (1969) have reiterated this conclusion.   This is misleading because it assumes a specific TRM close to the saturation magnetization value of magnetite (480 emu·cm^{-3}).   The correct approach, using NÉEL's (1949) theory, indicates that specific TRM's are generally much smaller (Fig. 1).   Nevertheless the effectiveness of SD grains as TRM carriers remains—particles just below the SD threshold and having axial ratio 1.5 to 1, claimed by STACEY and BANERJEE (1974, p. 109) to occur commonly in igneous rocks, have specific TRM $\sim$ 50 emu·cm^{-3} and thus need only occupy a volume fraction of 20 ppm to account for a typical basalt remanence of 10^{-3} emu·cm^{-3}.   By comparison 1 μm magnetite grains possess PSD thermoremanence of about 2 emu· cm^{-3} (DUNLOP, 1973) and MD particles ($>$20 μm) yield TRM's of 0.1 emu·cm^{-3} (STACEY and BANERJEE, 1974).

In conclusion it must be stressed that although the natural occurrence of SD particles has been demonstrated in some test cases, there is still no firm data on the actual size distributions and relative proportions of SP, SD, PSD and MD grains found in rocks. Where intergrown grains are present the SD 'Alnico model' seems appropriate, where isolated titanomagnetite grains are involved the SD 'tape recorder model' is satisfactory. Where pure Fe_3O_4 particles are involved the situation is less clear, and current thinking stresses the role of PSD moments (STACEY and BANERJEE, 1974; DUNLOP, 1977; BANERJEE, 1977). In all cases, including that of 'mixed' remanence, it should be kept in mind that progressive alternating field demagnetization, as undertaken in all modern paleomagnetic work, can be expected to enhance the single domain contribution.

Financial support was provided by the National Research Council of Canada.

REFERENCES

BANERJEE, S.K., On the origin of stable remanence in pseudo-single domain grains, *J. Geomag. Geoelectr.*, **29**, 319–329, 1977.

BATE, G., Oxides for magnetic recording, in *Magnetic Oxides*, edited by D.J. Craik, p. 689–742, John Wiley & Sons, New York, 1975.

BRECHER, A. and L. ALBRIGHT, The thermoremanence hypothesis and the origin of magnetization in iron meteorites, *J. Geomag. Geoelectr.*, **29**, 379–400, 1977.

BRECHER, A. and M. CUTRERA, A scanning electron microscope (SEM) study of the magnetic domain structure of iron meteorites and their synthetic analogues, *J. Geomag. Geoelectr.*, **28**, 31–45, 1976.

BUTLER, R.F. and S.K. BANERJEE, Theoretical single-domain grain size range in magnetite and titanomagnetite, *J. Geophys. Res.*, **80**, 4049–4058, 1975.

CHAMPNESS, P.E., Nucleation and growth of iron oxides in olivines (Mg, Fe)$_2$SiO$_4$, *Min. Mag.*, **37**, 790–800, 1970.

CREER, K.M. and N. PETERSEN, Thermochemical magnetization in basalts, *Z. Geophys.*, **35**, 501–515, 1969.

CULLITY, B.D., *Introduction to Magnetic Materials*, pp. 666, Addison-Wesley, Reading, Massachusetts, 1972.

DAVIS, P. and M.E. EVANS, Interacting Single-Domain Properties of Magnetite Intergrowths, *J. Geophys. Res.*, **81**, 989–994, 1976.

DE VOS, K.J., Alnico permanent magnet alloys, in *Magnetism and Metallurgy*, edited by A.E. Berkowitz and E. Kneller, p. 473–512, Academic Press, New York, 1969.

DUNLOP, D.J., Thermoremanent magnetization in submicroscopic magnetite, *J. Geophys. Res.*, **78**, 7602–7613, 1973.

DUNLOP, D.J., The hunting of the 'psark', *J. Geomag. Geoelectr.*, **29**, 293–318, 1977.

DUNLOP, D.J. and G.F. WEST, An experimental evaluation of single domain theories, *Rev. Geophys.*, **7**, 709–757, 1969.

EVANS, M.E. and M.W. McELHINNY, An investigation of the origin of stable remanence in magnetite-bearing igneous rocks, *J. Geomag. Geoelectr.*, **21**, 757–773, 1969.

EVANS, M.E. and M.L. WAYMAN, An investigation of small magnetic particles by means of electron microscopy, *Earth Planet. Sci. Lett.*, **9**, 365–370, 1970.

EVANS, M.E. and M.L. WAYMAN, An investigation of the role of ultra-fine titanomagnetite intergrowths in paleomagnetism, *Geophys. J.R. Astron. Soc.*, **36**, 1–10, 1974.

EVANS, M.E., M.W. McELHINNY, and A.C. GIFFORD, Single domain magnetite and high coercivities in a gabbroic intrusion, *Earth Planet. Sci. Lett.*, **4**, 142–146, 1968.

EVDOKIMOV, V.B., The magnetic interaction of superparamagnetic particles, *Russ. J. Phys. Chem.*, **37**, 1018–1019, 1963.

GRAHAM, J.W., Changes in ferromagnetic minerals and their bearing on magnetic properties of rocks, *J. Geophys. Res.*, **58**, 243–260, 1953.

HADFIELD, D. (ed.), *Permanent Magnets and Magnetism*, pp. 556, Iliffe Books, London, 1962.

HARGRAVES, R.B. and W.M. YOUNG, Source of stable remanent magnetism in Lambertville diabase, *Am. J. Sci.*, **267**, 1161–1177, 1969.

HOBLITT, R.P. and E.E. LARSON, New combination of techniques for determination of the ultrafine structure of magnetic minerals, *Geology*, **3**, 723–726, 1975.

HOYE, G.S. and M.E. EVANS, Remanent magnetizations in oxidized olivines, *Geophys. J.R. Astron. Soc.*, **41**, 139–151, 1975.

HOYE, G.S. and W. O'REILLY, Low temperature oxidation of ferro-magnesian olivines—a gravimetric and magnetic study, *Geophys. J.R. Astron. Soc.*, **33**, 81–92, 1973.

JENSEN, S.D. and P.N. SHIVE, Cation distribution in sintered titanomagnetites, *J. Geophys. Res.*, **78**, 8474–8480, 1973.

KOBAYASHI, K., An experimental demonstration of the production of chemical magnetization with Cu–Co alloy, *J. Geomag. Geoelectr.*, **12**, 148–163, 1961.

LARSON, E., M. OZIMA, T. NAGATA, and D. STRANGWAY, Stability of remanent magnetization of rocks, *Geophys. J.R. Astron. Soc.*, **17**, 263–292, 1969.

MANSON, A.J. and W. O'REILLY, Submicroscopic texture in titanomagnetite grains in basalt studied using the torque magnetometer and electron microscope, *Phys. Earth Planet. Inter.*, **11**, 173–183, 1976.

MORRISH, A.H. and L.A.K. WATT, Effect of the interaction between magnetic particles of the critical single-domain size, *Phys. Rev.*, **105**, 1476–1478, 1957.

MURTHY, G.S., M.E. EVANS, and D.I. GOUGH, Evidence of single-domain magnetite in the Michikamau Anorthosite, *Can. J. Earth Sci.*, **8**, 361–370, 1971.

NÉEL, L., Theorie du trainage magnetique des ferromagnetiques en grains fins avec application aux terres cuites, *Ann. Geophys.*, **5**, 99–136, 1949.

NÉEL, L., Some theoretical aspects of rock magnetism, *Adv. Phys.*, **4**, 191–242, 1955.

PAUTHENET, R., Aimantation Spontanée des Ferrites, *Ann. Phys.*, **7**, 710–747, 1952.

POWELL, D.W., Significance of differences in magnetization along certain dolerite dykes, *Nature*, **199**, 674–676, 1963.

RADHAKRISHNAMURTHY, C. and E.R. DEUTSCH, Magnetic techniques for ascertaining the nature of iron oxide grains in basalts, *J. Geophys.*, **40**, 453–465, 1974.

RADHAKRISHNAMURTHY, C., N.P. SASTRY, and E.R. DEUTSCH, Ferromagnetic behaviour of interacting superparamagnetic particle aggregates in basaltic rocks, *Pramana*, **1**, 61–65, 1973.

RYALL, P.J.C. and J.M. ADE-HALL, Radial variation of magnetic properties in submarine pillow basalt, *Can. J. Earth Sci.*, **12**, 1959–1969, 1975.

SOFFEL, H.C., Die Beobachtung von Weisschen Bezirken auf einem Titanomagnetitkorn mit einem Durchmesser von 10 Mikron in einem Basalt, *Z. Geophys.*, **34**, 175–181, 1968.

SOFFEL, H.C., The origin of thermoremanent magnetization of two basalts containing homogeneous single phase titanomagnetite, *Earth Planet. Sci. Lett.*, **7**, 201–208, 1969.

SOFFEL, H.C., The single-domain—multidomain transition in natural intermediate titanomagnetites, *Z. Geophys.*, **37**, 451–570, 1971.

STACEY, F.D., Theory of the magnetic properties of igneous rocks in alternating Fields, *Philos. Mag.*, **6**, 1241–1260, 1961.

STACEY, F.D. and S.K. BANERJEE, *The Physical Principles of Rock Magnetism*, pp. 195, Elsevier, Amsterdam, 1974.

STRANGWAY, D.W., Magnetic properties of diabase dikes, *J. Geophys. Res.*, **66**, 3021–3031, 1961.

STRANGWAY, D.W., E.E. LALSON, and M. GOLDSTEIN, A possible cause of high magnetic stability in volcanic rocks, *J. Geophys. Res.*, **12**, 3787–3795, 1968.

WILSON, R.L. and N.D. WATKINS, Correlation of magnetic polarity and petrological properties in Columbia Plateau basalts, *Geophys. J.R. Astron. Soc.*, **12**, 405–424, 1967.

Adv. Earth Planet. Sci., **1**, 45–52, 1977

Domain Structure of Titanomagnetites and Its Variation
with Temperature

H.C. SOFFEL

Institut für Allgemeine und Angewandte Geophysik, Universität München,
Theresienstrasse, München, Germany

(Received June 20, 1977)

In a Tertiary lava flow with optically apparently homogeneous titanomagnetites (Curie temperatures less than 100°C) a range of Curie temperatures from less than −40°C in the central to more than 70°C in the marginal zones of the crystals was found by observing the variation of the domain structure with temperature. The effect is explained by the different degree of maghemitization. TRM production in fields of 35 Oe revealed that the direction of the external field had little influence on the magnetic domain pattern.

1. Introduction

Thermoremanent magnetization (TRM) is produced in rocks during cooling in a magnetic field from high temperatures above the Curie temperature down to low temperature (about room temperature). The properties of TRM have been summarized recently by STACEY and BANERJEE (1974). Above the Curie temperature all ore grains are paramagnetic and have no domain configuration. By cooling below the Curie temperature saturation magnetization increases rapidly from practically zero to such high values that the energy balance of the crystal is strongly affected. In order to achieve a state of minimum energy, magnetic domains are developed. External fields, crystal anisotropy intensity of saturation magnetization, internal and external uniaxial stresses as well as the shape and irregularities of the external and internal surfaces of the crystal have an influence on the geometry, spatial distribution and size of the magnetic domains (see also STACEY and BANERJEE, 1974, p. 41). Each domain is magnetized more or less to saturation along an axis of easy magnetization. In the absence of an external field and of uniaxial stresses the domain configuration tends to be such that the net magnetic moment of the crystal is zero. External fields are believed to influence the development of magnetic domains in such a way that a remanence is produced. The sum of remanences of an aggregate of randomly oriented ore grains is in general assumed to be parallel to the direction of the external field. However the low intensity of TRM of true multidomain particles produced in a field of about 0.5 Oe in comparison with the intensity of saturation remanence already indicates that small fields (say less than 1 Oe) seem to have little influence on the domain structure. It is generally believed that small external fields influence the domain structure mainly at or immediately below Curie temperature in such a way that the initially formed domains are magnetized in the easy magnetization direction which is closest to the direction of the external field.

The domain structure of natural titanomagnetites in basalts and its variation with temperature and direction of the external field during production of TRM was studied under the above mentioned aspects. The aim was to see how the domains develop in a small ore grain during cooling below Curie temperature.

2. Rock Magnetic Properties of the Investigated Basalt

For this special study a Tertiary basalt sample from the Southern Alps (flow No. 24/2, Soffel, 1975) was selected which contained titanomagnetites of apparently low oxidation state with bulk Curie temperatures of 100°C or less (Fig. 1). It was necessary to choose very low Curie temperature minerals because the sample could not be heated much above 100°C while under the microscope for technical reasons. Other rock magnetic properties of this lava flow have been published elsewhere (Soffel 1975). The ore grains appear to be optically homogeneous up to very large magnifications (1,300 times) before and after etching and show no signs of either deuteric oxidation or low temperature oxidation due to weathering. The bulk coercive force is about 70 Oe.

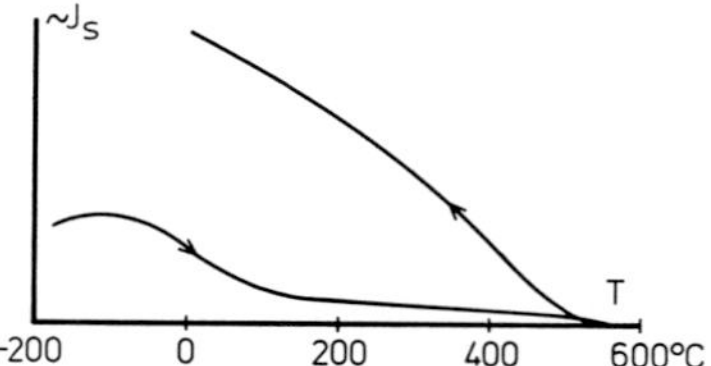

Fig. 1. Variation of saturation magnetization (measured in 5,000 Oe) with temperature between −170° and +600°C. Arrows indicate heating and cooling cycle. The increased intensities of the cooling cycle are due to exsolution of the titanomagnetites at high temperatures.

3. Observation of Magnetic Domains

It has been shown (Soffel, 1968, 1971) that the Bitter pattern technique is an adequate tool for the observation of magnetic domains of titanomagnetites in a rock matrix. Domain walls are observed on the polished surface of the crystals as dark lines. They represent the domain structure near the surface. In very small ore grains the observed domain structure on the surface can be regarded as representative for the interior of the crystal as well. Special care must be taken to get strainfree surfaces. This can be achieved by first mechanical and afterwards ionic polishing (Soffel and Petersen 1971). For the domain observation at various temperatures, an ester-based magnetite colloid ('Ferrofluid') was used instead of the magnetite colloid after Elmore (1938). Temperature could be varied between about −50°C

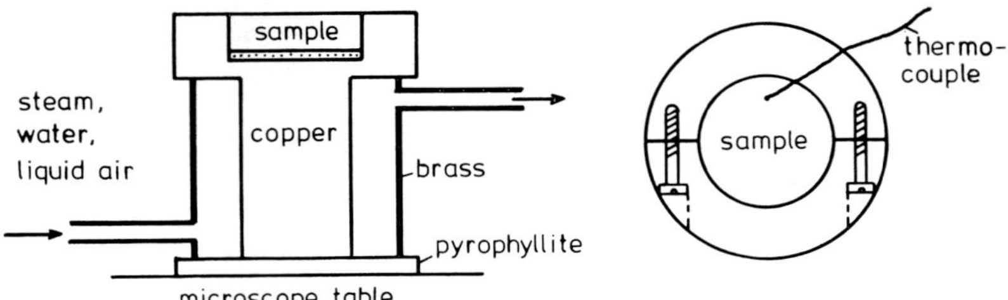

Fig. 2.   Non-magnetic device for heating and cooling of polished sections while under the microscope. The cylindrical samples have a diameter of 2.5 cm and a height of about 1 cm.

(freezing point of the immersion oil) and about $+100°C$ (highest temperature for the microscope objectives). Heating and cooling was done with the device shown in Fig. 2 by pumping liquid air, hot water or steam around the copper block. A thermocouple clamped to the sample surface was used to measure the temperature of the ore grains. External fields parallel to the surface of the polished section were produced with a small electromagnet. The field could be varied between ± 35 Oe which is about 100 times the intensity of the horizontal component of the earth's field in the laboratory.

4.   Domains at Various Temperatures

It could be concluded from the variation of saturation magnetization with temperature (Fig. 1) that only a certain fraction of the ore grains is ferrimagnetic at room temperature and should show domain structure (group I). The rest of the ore grains (group II) are supposed to have Curie temperatures below room temperature. They are paramagnetic and without domain configuration. Due to the optical homogeneity, all the ore grains should belong to either group I or II. However it was surprising to see that only the smallest ore grains with a diameter of less than about 6 microns were composed entirely of ferrimagnetic material and showed magnetic domains. The larger ore grains were only partly ferrimagnetic at room temperature, with large parts still paramagnetic, or they belonged to group II (no domain structure at all).

Figure 3a–3d shows a large ore grain with a diameter of about 60 microns at various temperatures. At room temperature (Fig. 3a) most of the crystal is above its Curie temperature with the exception of some small marginal zones which show a complicated domain structure. Details of this configuration shall not be discussed at the moment. At 70°C (Fig. 3b) most of these regions have become paramagnetic as well and lost their domain structure. After cooling to room temperature in a 35 Oe field ($\rightarrow$), the domain structure (Fig. 3c) is not quite the same as before (see Fig. 3a) but the same zones are ferrimagnetic again. At $-40°C$, (Fig. 3d) most but not all of the crystal has become ferrimagnetic. A small grain, which showed no domains at 20°C and higher temperaturers, has become ferrimagnetic as well at $-40°C$ with a domain

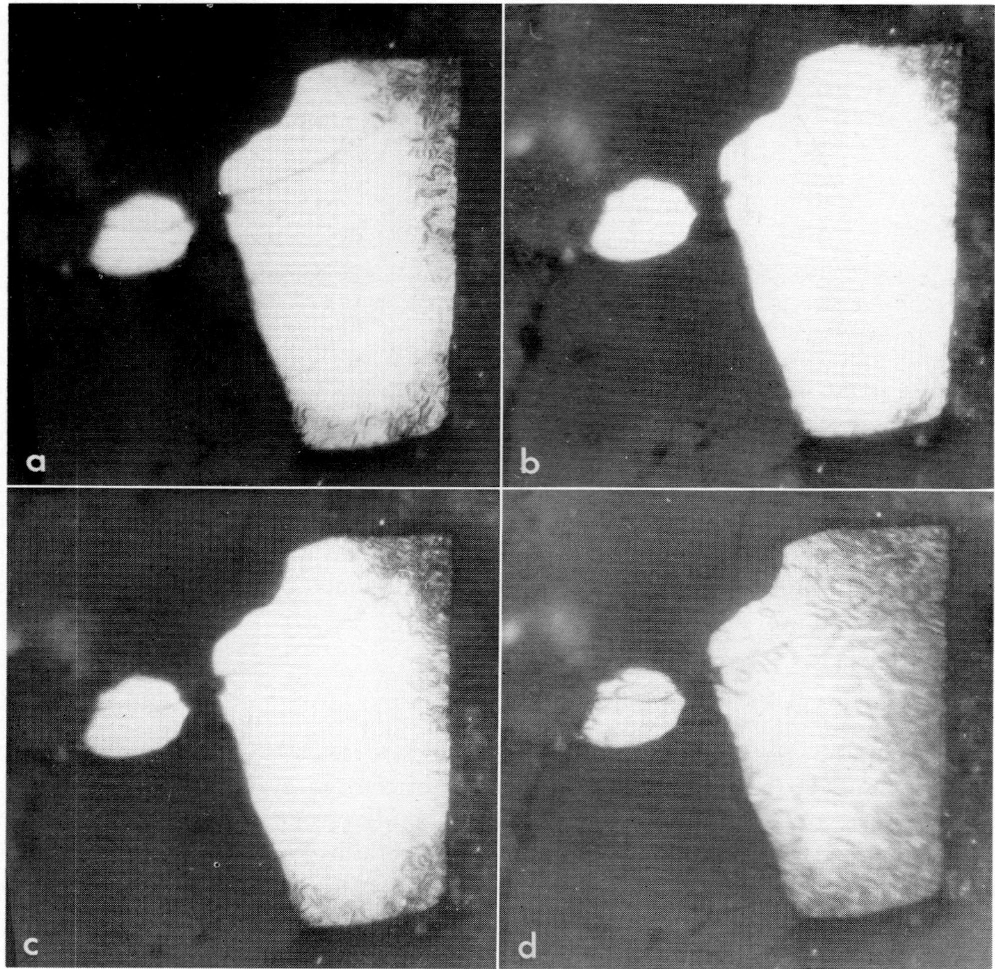

Fig. 3.   Ore grain (diameter about 60 microns) at various temperatures. (a) room temperature, (b) 70°C, (c) room temperature, (d) −40°C. For details, see text.

structure. The consequences of this range of Curie temperatures, from less than −40°C to more than +70°C, in an optically apparently homogeneous crystal for TRM is discussed in the next section.

Figure 4a–4d shows a smaller titanomagnetite grain (length about 25 microns) which is almost entirely ferrimagnetic at room temperature (Fig. 4a). At 85°C, the grain is paramagnetic and has lost its domain configuration (Fig. 4b). Cooling to room temperature in 35 Oe parallel to the surface (↓) reproduces substantially the original domain configuration (Fig. 4c). When the experiment is repeated with another direction of the external field of 35 Oe (→), the domain configuration is still not drastically altered (Fig. 4d). A schematic drawing of Fig. 4a–4d is shown in Fig. 5. Dark lines are the observed domain wall traces.

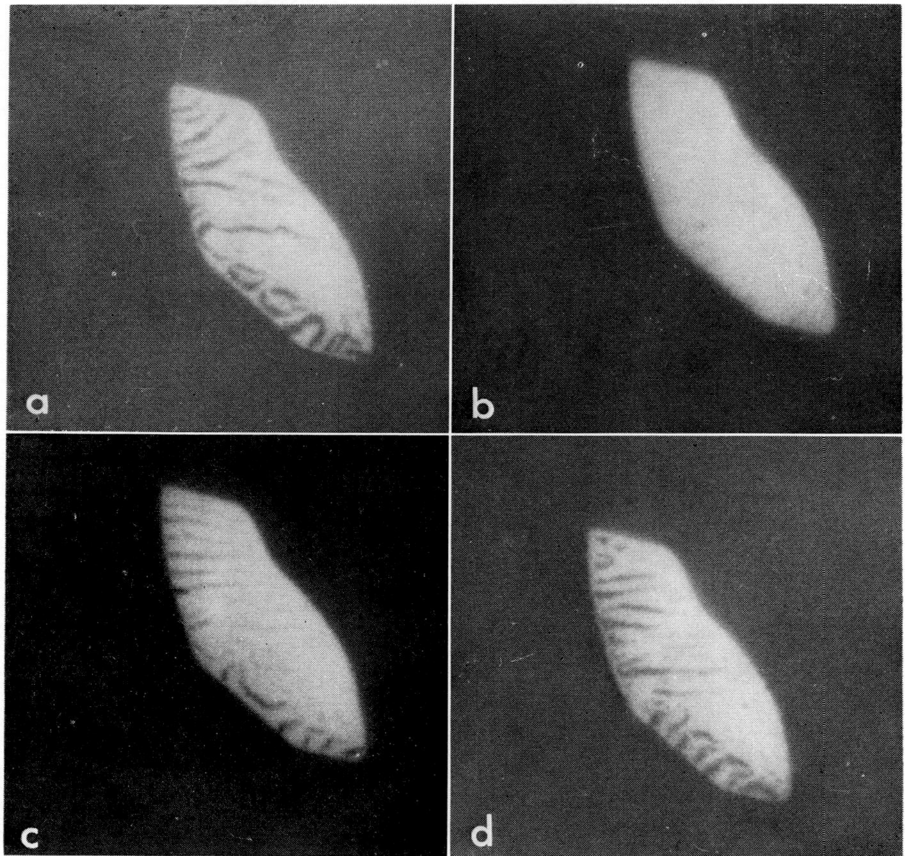

Fig. 4.   Ore grain (length about 25 microns) at various temperatures.   (a) room temperature,
(b) 85°C, (c) and (d) room temperature.   For details, see text.

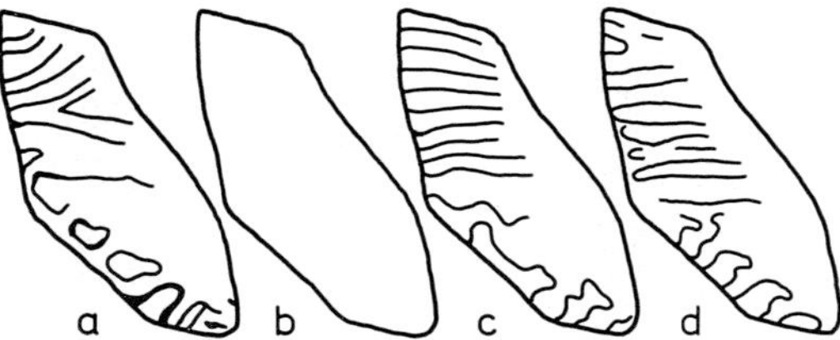

Fig. 5.   Schematic drawing of the domain configurations shown in
Fig. 4a–d.

Figure 6a–6f shows another small titanomagnetite grain (length about 10 microns).   Figure 6a represents the original domain configuration at room temperature, Figure 6b shows the paramagnetic ore grain at 85°C.   After cooling in 35 Oe parallel to the surface (→), almost the original configuration reappears (Fig. 6c). Reheating above Curie temperature and cooling in 35 Oe with a different horizontal

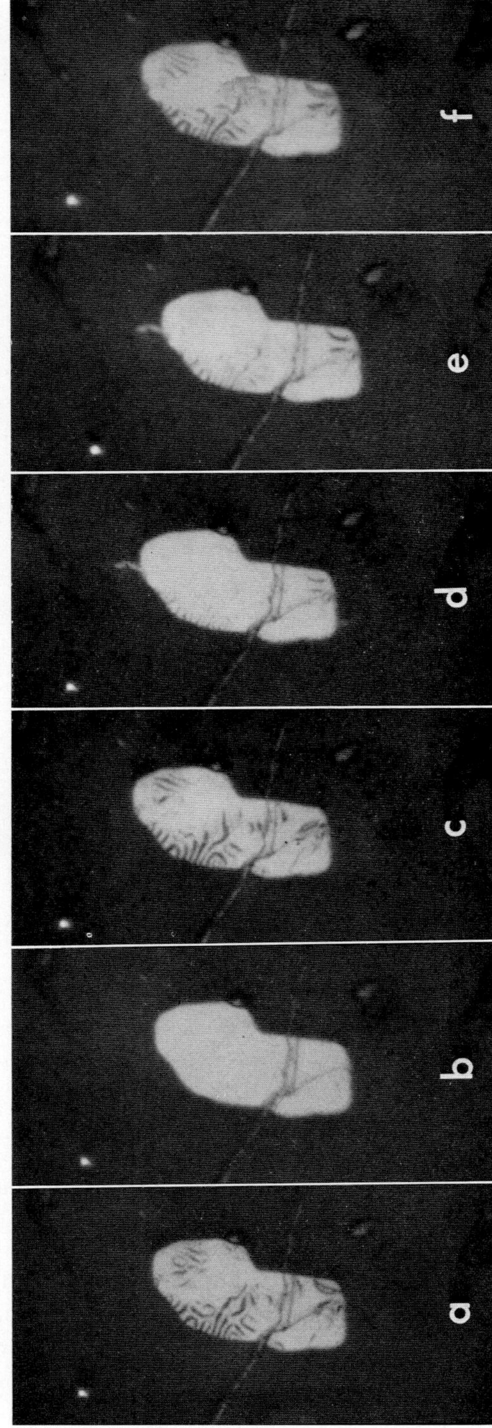

Fig. 6. Ore grain (length about 10 microns) at various temperatures (a) room temperature, (b) 85°C, (c) room temperature, (d) 60°C, (e) 40°C, (f) 18°C. For details, see text.

direction (↓) shows in Fig. 6d the configuration at 60°C, in Fig. 6e at 40°C and in Fig. 6f at 18°C. The configurations of Fig. 6a, 6c and 6f, though generated in different field directions, vary only in details in the upper right and central part of the crystal. It is difficult to estimate whether these differences are sufficient to produce differences in the net magnetic moment of the ore grain. On the other hand, the observed TRM in a rock must be composed of such individual effects in the ore grains. However, effects in a single ore grain may not necessarily be representative for a rock sample as a whole.

Figures 4–6 contain some surprising results. (i) Domain walls are not always parallel and straight. (ii) The domain magnetization is not always parallel to the long axis of the ore grain. (iii) The production of a TRM in a field one hundred times the earth's magnetic field does not influence the domain configuration very much.

Observations (i) and (ii) are common with titanomagnetites of low saturation magnetization. In pure magnetite, the domain walls tend to be straight and also the shape anisotropy influences the domain configuration in the expected way. The large intensity of saturation magnetization in combination with the moderate intensities of crystal anisotropy and magnetostrictive constants, K and λ respectively, prevent or reduce the occurrence of 'free magnetic charges' along curved domain walls due to local inhomogeneities and stress fields within the crystal or along the margins. This situation does not hold in low Curie temperature titanomagnetites with low values of saturation magnetization and large values of K and λ (SYONO 1965). In titanomagnetites crystal anisotropy and magnetostrictive effects are more important and curved domain walls are often formed in order to reduce the energy of the ore grain at the expense of some additional magnetostatic energy along curved domain walls and along the margins of the crystal.

The most surprising observation to the author was the slight influence of rather strong external fields on the domain configuration during TRM production. Three possible explanations can be considered. (i) Although the domain configuration disappeared at a certain temperature on the visible part of a grain, it may still be present in other invisible parts of the ore grain. (ii) The Curie temperature was not completely attained but the saturation magnetization and the stray fields along the domain wall traces have become so small that they could no longer attract any particles of the magnetite colloid. (iii) The internal stress fields of the ore grains are so large that they have a stronger influence on the domain configuration than the external field applied (35 Oe). In order to check the different possibilities, experiments with higher temperatures, larger external fields and various uniaxial stresses are planned.

5.   Discussion and Conclusions

The very extended range of Curie temperature within an ore grain is believed to be due to maghemitization, with the highest degree of alteration along the margins and cracks. Such maghemitization does not affect the titanium-iron-ratio essentially and can therefore not be observed easily in microprobe studies. In X-ray analysis, it cannot be distinguished whether a certain diffraction line width is due to homogeneous grains with variable chemical composition or due to inhomogeneous ore grains. This may be the reason why the large range of Curie temperatures within the titanomagnetite grains was not detected earlier. The effect does not seem to be unique for the studied lava flow No. 24/2. Other basalts of the same region with very low Curie temperatures showed the same effect.

During the implacement of a volcanic rock the increased access of oxygen may not always lead to such a high oxidation state of titanomagnetites that exsolution into two different phases takes place. Lower oxidation states in the form of non stoichiometric titanomagnetites (titanomaghemites) may also occur, with the highest degree of maghemitization in the outer parts of the ore grains.

All theories about the origin of TRM assume a ferrimagnetic ore phase with a single Curie temperature or two separated phases with different Curie temperatures. These models seem to be realistic for rocks with magnetite or completely exsolved titanomagnetites, but not for rocks with primary non-stoichiometric titanomagnetites with decreasing Curie temperature from the margins towards the centre of the ore grain. It was shown in Fig. 3 how the domain configuration 'grows' steadily from the ferrimagnetic zones into the previously paramagnetic zones of the crystal during cooling. Their TRM is therefore eventually more influenced by the magnetic domains of the higher Curie temperature zones and their TRM information than by the weak external earth's magnetic field. This puts some question marks on the law of additivity of partial thermoremanent magnetizations on a microscopic scale in rocks with non-stoichiometric titanomagnetites. Figures 4–6 also show that the decay of intensity of remanence of basalts during thermal demagnetization may be only partly due to a spectrum of blocking temperature. A range of Curie temperatures, which can be as much as 100°C, must also be considered.

The author is very much indebted to Dr. D. Dunlop for reading the paper at the fall 1976 meeting of AGU and for his helpful suggestions. Special thanks are due to the Munich rock magnetism group (Drs. Angenheister, Schult, Petersen, Pohl, Schmidbauer) for many useful discussions. The financial help of the Deutsche Forschungsgemeinschaft is gratefully acknowledged.

REFERENCES

ELMORE, W.C., Ferromagnetic colloid for studying magnetic structures, *Phys. Rev.*, **54**, 309–310, 1938.

SOFFEL, H., Die Bereichsstrukturen der Titanomagnetite in zwei tertiären Basalten und die Beziehung zu makroskopisch gemessenen magnetischen Eigenschaften dieser Gesteine, Habilitation Thesis, Fac. Nat. Sciences, University of Munich, Germany, 1–142, 1968.

SOFFEL, The single domain-multidomain transition in natural intermediate titanomagnetites, *Z. Geophys.*, **37**, 451–470, 1971.

SOFFEL, H., Rock magnetism of the Monti Lessini and Monte Berici volcanites and age of volcanism deduced from the Heirtzler Polarity time Scale, *J. Geophys.*, **41**, 401–411, 1975.

SOFFEL, H. and N. PETERSEN, Ionic etching of titanomagnetite grains in basalts. *Earth Planet. Sci. Lett.*, **11**, 312–316, 1971.

STACEY, F.D. and S.K. BANERJEE, *The Physical Principles of Rock Magnetism*, Elsevier Scientific Publishing Company, Amsterdam, London, New York, 1974.

SYONO, Y., Magnetocrystalline anisotropy and magnetostriction of Fe_3O_4-Fe_2TiO_4—series with special application to rock magnetism, *Jpn. J. Geophys.*, **4**, 71–143, 1965.

Adv. Earth Planet. Sci., **1**, 53–60, 1977

The Demagnetization Field of Multidomain Grains

Ronald T. Merrill[*]

*Geophysics Program and Oceanography Department,
University of Washington, Seattle, U.S.A.*

(Received June 20, 1977)

The demagnetization 'constant', N, enters into all theoretical considerations of multidomain remanence. Although most workers believe N to be related simply to the shape of the grain, no theoretical study of N has previously been completed. It is shown in this paper that N monotonically increases with increase in remanence in multidomain grains. Although N can be approximated by a constant in the limit of small displacements of domain walls from their demagnetized sites, this constant critically depends on the domain geometry.

1. Introduction

The successes of paleomagnetism are based on the remarkable observation that many rocks retain a record of the direction of magnetization acquired hundreds to billions of years previously. The important works of NÉEL on single domain theory (1949) and on multidomain theory (1955) remain at the foundation of most theoretical work which attempts to explain the remarkable stability of thermal remanent magnetization (TRM). The successes of Néel's theories are impressive. However, in recent years it has become increasingly clear that Néel's theories require modifications (for example DUNLOP and WEST, 1969; STACEY and BANERJEE, 1974; DUNLOP and WADDINGTON, 1975; SCHMIDT, 1976). One particular problem that has emerged can be illustrated by considering the mineral magnetite. It appears that the stable single domain size range for equant magnetite grains is very limited (BUTLER and BANERJEE, 1975). Yet larger grains appear to possess some properties that are more expected of single domain grains than multidomain grains (VERHOOGEN, 1959). Verhoogen suggests that the solution to this problem is that magnetization in screw dislocations in multidomain material acts like single domain grains. Although the details of VERHOOGEN's model (1959) appear unsatisfactory (SHIVE, 1969), the basic idea that 'something' inside large grains behaves in a single domain manner remains as a viable construct; this behavior is often referred to as pseudo-single domain behavior (STACEY, 1963; STACEY and BANERJEE, 1974). An alternative suggestion has been advanced by SCHMIDT (1973), who argues that two-domain grains behave very much like Néel's single domain grains, and very differently from much larger multidomain grains.

Therefore, it seems timely to reconsider the various assumptions of multidomain

---

[*] Contribution Number 973—Department of Oceanography.

theory. This is done in the next section. Subsequently, one of these assumptions, that dealing with the demagnetization energy, is considered in detail.

2. The Assumptions of Multidomain Theory

At the onset we distinguish between 'true' multidomain theories and pseudo single domain theories. 'True' multidomain theories have been reviewed by DUNLOP and WADDINGTON (1975); the critical distinction being that 'true' multidomain grains acquire their remanence through domain wall displacements. Pseudo single domain theories deal with nonuniformly magnetized grains which alledgedly acquire their stable remanence by mechanisms other than domain wall displacement. (For example, STACEY and BANERJEE, 1974, suggest that the remanence resides inside domain walls.) Hereafter, the modifier 'true' will be dropped, since its use might connote unintended implications of validity.

The usual assumptions made to obtain the magnetization, J, in multidomain grains are as follows: (1) The remanence is acquired and lost through domain wall motion: (2) Domain walls can be treated as being infinitesimally thick; (3) Equilibrium occurs in the presence of an external field, h, when the magnetic energy, $\frac{1}{2}\bar{J}\cdot\bar{h}$, is balanced by the demagnetization energy; (4) A remanence is acquired when domain walls are blocked by energy barriers, which are thought to be associated with crystal defects; (5) These energy barriers are associated with a microscopic coercivity, H_c, whose temperature dependence can be expressed as βJ_S^n, where β and n are positive constants and J_S is the saturation magnetization. An additional assumption is required when dealing with TRM: (6) The basic domain configuration is not strongly temperature dependent. That is, a grain with p domains at elevated temperatures will still have p domains at room temperature.

Many of these assumptions have been challenged. However, somewhat surprisingly, the nature of the demagnetization field has never been very critically examined. In all prior multidomain theories, the demagnetization field has been taken to be $-NJ$, where N is called the demagnetization factor, in analogy with single domain theory. This paper deals only with this assumption.

We briefly review multidomain theory to see how the demagnetization field assumption is used. This development is also useful later when we consider modified multidomain theories.

DUNLOP and WADDINGTON (1975) start with the following equations, which are consistent with the above assumptions. Equilibrium predicts that

$$h - NJ = -H_c = -\beta J_S^n \qquad (1)$$

or

$$J = \frac{h + \beta J_S^n}{N}. \qquad (2)$$

Blocking of TRM can be defined in terms of j, which is defined by the equation:

$$j \equiv \frac{J}{J_S}. \qquad (3)$$

Blocking occurs when there is no domain wall movement with decrease in temperature. That is,

$$\frac{dj}{dT} = 0 . \tag{4}$$

From Eqs. (1) and (4), we obtain:

$$h = (n-1)H_c . \tag{5}$$

Once blocking has occurred, the TRM increases only as the saturation magnetization increases. This can be taken into account by multiplying Eq. (2) by J_{So}/J_{SB}, the ratio of saturation magnetization at room temperature to that at blocking temperature. Let $H_{co} \equiv$ microscopic coercivity at room temperature $= \beta J_{So}^n$ and $H_{cB} \equiv$ microscopic coercivity at blocking temperature $= \beta J_{SB}^n$. The Eq. (5) and the modified Eq. (2) can be used to obtain:

$$J = \frac{n}{(n-1)^{(1-1/n)}} \frac{h^{1-1/n} H_{co}^{1/n}}{N} \tag{6}$$

as given by Dunlop and Waddington (1975).

There are two important results of Eq. (6). The first is that TRM is not linearly proportional to the external field. Although it is not clear what value of n should be used, essentially all workers agree that n is greater than 1 and less than 9. The second result is that the intensity of TRM is inversely proportional to N.

3.   The Demagnetization Factor

The first question that must be faced is: what is N in Eq. (6)? Most workers believe N to be related to the shape of the grain, in analogy with single domain theory. However, it is important to realize that N is only well defined for single domain grains of certain simple shapes, such as ellipsoids of revolution. For non-uniformly magnetized grains, the demagnetization field will vary from location to location inside the grain. What is N in such a case?

Insight into how one might define N comes from the important work of Kittel (1949), as reiterated by Chikazumi (1964, p. 209–211). Figure 1 shows the demag-

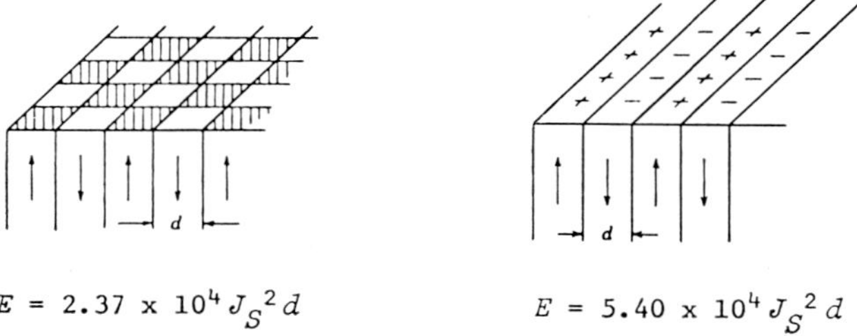

Fig. 1.   The internal magnetic energy associated with bound surface charges for the checkerboard (left) and sheeted (right) domain structures. The examples given are for completely demagnetized states.

netization energy associated with checkerboard and sheeted domain structure. Those results suggest that the so called 'demagnetization factor' might depend critically on the domain structure and not on the grain's shape. Note, however, that Kittel and Chikazumi have only treated domain structures in which there is no net magnetization. Therefore, no term associated with the particular demagnetization energies given in Fig. 1 appears in Eq. (1). We break the demagnetization energy up into two parts, E_o and E_c. E_o is the demagnetization energy associated with a completely demagnetized multidomain grain (examples are given in Fig. 1). E_c is the extra demagnetization energy that results when a grain carries a remanence, that is, the extra energy that results from domain wall displacements from the zero field equilibrium position. It is the demagnetization field associated with this energy that should appear in Eq. (1). We define N by the equation:

$$E_c = \frac{1}{2} N J^2$$

or

(7)

$$N = \frac{2E_c}{J^2}.$$

It is important to realize that, if we use the N of Eq. (7) in Eq. (1), then we are still making a rather critical assumption. We are assuming that it is the average demagnetization energy for the entire grain that is important in TRM. This might not be the case if certain regions of a multidomain grain are blocked before others. In the latter situation, localized demagnetization energies must be considered. However, in the attempt to obtain a first order approximation for N, we neglect this last possibility.

We now set out to show that N depends critically on the domain structure of a multidomain grain and that it increases with increase in mean wall displacement from the zero-field equilibrium position. We do this by considering the sheeted domain structure shown in Fig. 2. Several other configurations can be easily calculated and such calculations lead to changes in detail but not in principle. The important work of Kittel (1949) and Chikazumi (1964, p. 209–211) provide the necessary insight into the type of calculation required to determine N.

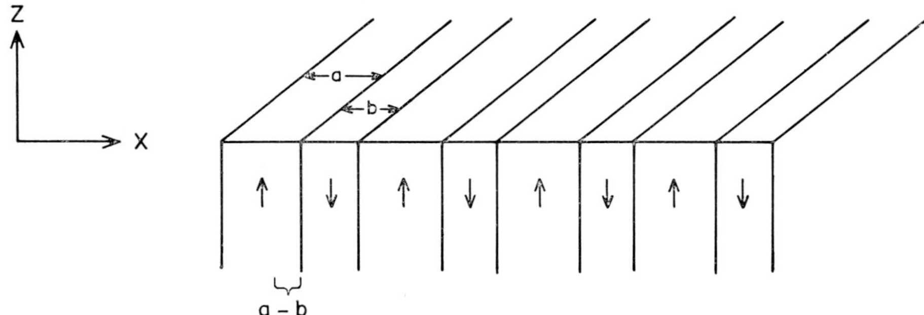

Fig. 2. Sheeted domain structure which exhibits a net remanence. This remanence is a result of displacement of every other domain wall by a distance $a-b$ from the completely demagnetized state.

External to the grain, $\bar{V} \times \bar{h} = 0 = \bar{V} \cdot \bar{h}$. This implies that there exists a potential ϕ, such that Laplace's equation holds:

$$\nabla^2 \phi = 0 \, . \tag{8}$$

Now, using MKS units, the magnetic induction, $\bar{B}$, is given by $\bar{B} = \bar{\bar{\mu}}_0(\bar{h} + \bar{J})$, where $\bar{\bar{\mu}}_0$ is the magnetic permeability in air. Using the fact that $\bar{V} \cdot \bar{B} = 0$ everywhere, and that the magnetic bound charge is given by $\rho = \bar{V} \cdot \bar{J}_S$, one has the boundary conditions:

$$\left(\frac{\partial \phi}{\partial z}\right)_{z \to +0} = \frac{-\rho}{2\mu_0} \quad 0 < x < a \, , \tag{9}$$

$$\left(\frac{\partial \phi}{\partial z}\right)_{z \to +0} = \frac{\rho}{2\mu_0} \quad a < x < a+b \, . \tag{10}$$

The following additional boundary conditions are also easily seen to be valid for the case under consideration:

$$\lim_{z \to \infty} \phi(z) = 0 \tag{11}$$

$$\phi[x + (a+b)L] = \phi(x) \quad \text{for } L = \text{integer.} \tag{12}$$

Eqs. (8) through (12) constitute a boundary value problem that is easily solved by separation of variables. The solution is:

$$\phi = A_0 + \sum_{m=1}^{\infty} A_m \cos\left(\frac{m\pi x}{a+b}\right) e^{-\frac{m\pi z}{a+b}} + \sum_{m=1}^{\infty} B_m \sin\left(\frac{m\pi x}{a+b}\right) e^{-\frac{m\pi z}{a+b}} \tag{13}$$

where

$$A_m = \frac{\rho(a+b)}{(m\pi)^2 \mu_0} \sin\left(\frac{m\pi a}{a+b}\right) \tag{14}$$

$$B_m = \frac{\rho(a+b)}{(m\pi)^2 \mu_0} \cos\left(\frac{m\pi a}{a+b}\right) \, . \tag{15}$$

A_0 is a constant of little consequence in this problem and may be dropped (ϕ uniquely defines the magnetic problem, external to the grain, within an arbitrary constant).

The next step in the calculation utilizes the important observation of KITTEL (1949) that one can obtain the self-demagnetization energy by calculating the surface energy. The surface energy, per unit area is:

$$E = \frac{1}{2}\rho\phi_{z=0} \, . \tag{16}$$

It is easy to show, by using $\bar{V} \cdot \bar{J}_S = \rho$ and a 'pill-box' approach at the boundary $z = 0$, that the surface charge per unit area is equal to J_S for the domain configuration of Fig. 2. Therefore, the average self-demagnetization energy per unit area is:

$$E = \sum_{m=1}^{\infty} \frac{J_S^2}{2\mu_0} \frac{(a+b)}{(m\pi)^2} \sin\left(\frac{m\pi a}{a+b}\right)\left[\frac{1}{a}\int_0^a \cos\left(\frac{m\pi x}{a+b}\right)dx - \frac{1}{b}\int_a^{a+b} \cos\left(\frac{m\pi x}{a+b}\right)dx\right]$$

$$+ \sum_{m=1}^{\infty} \frac{J_S^2}{2\mu_0} \frac{(a+b)}{(m\pi)^2} \cos\left(\frac{m\pi a}{a+b}\right)\left[\frac{1}{a}\int_0^a \sin\left(\frac{m\pi x}{a+b}\right)dx - \frac{1}{b}\int_a^{a+b} \sin\left(\frac{m\pi x}{a+b}\right)dx\right] \tag{17}$$

or

$$E=\sum_{m=1}^{\infty}\left\{\frac{J_S^2}{2ab}\frac{(a+b)^3}{\mu_0(m\pi)^3}\sin^2\left(\frac{m\pi a}{a+b}\right)-\frac{J_S^2(a+b)^2}{2ab\mu_0(m\pi)^3}\right.$$
$$\left.\times\left[(a+b)\cos^2\left(\frac{m\pi a}{a+b}\right)+\frac{(b-a[-1]^m)}{ab}\cos\left(\frac{m\pi a}{a+b}\right)\right]\right\}. \tag{18}$$

It is clear from Eq. (18) that the average internal self-demagnetization energy is dependent on the domain configuration. Following the examples provided by the earlier theories for multidomain grains we would like to express the self-demagnetization energy in terms of the remanent magnetization, J, which is linearly proportional to $J_S(a-b)$ (see Fig. 2). If any of the previous multidomain theories were correct, then the energy should be given by $\frac{1}{2}NJ^2$, where N depends solely on grain shape. As seen from (18), this is not the case. To see this more clearly, and to obtain simpler relationships to work with, we consider two extreme limits for Eq. (18).

We consider first the case where there is only a small displacement of the domain walls from the zero-field equilibrium position. This corresponds to the case of small remanence in multidomain grains. In this case, $a/(a+b)\simeq\frac{1}{2}$, etc. and (18) reduces to

$$E\simeq\frac{J_S^2(a+b)^3}{2\mu_0\pi^3 ab}\sum_{m\ \mathrm{odd}}\frac{1}{m^3}. \tag{19}$$

We further reduce Eq. (19), by noting that a and b are not independent of each other. We write:

$$a\equiv d+\varepsilon$$
$$b\equiv d-\varepsilon$$

where ε is the distance the domain wall is displaced and d is the zero-field equilibrium dimension of the domain (see Figs. 1 and 2). Therefore $(a+b)=2d$. $1/ab$ can be approximated by the first few terms in a Taylor's expansion for small ε:

$$\frac{1}{ab}=\frac{1}{d^2-\varepsilon^2}\simeq\frac{1}{d^2}+\frac{2\varepsilon^2}{d^4}+\frac{12\varepsilon^4}{d^6}. \tag{20}$$

We note that the displacement ε is linearly proportional to the remanence (The remanence increases as 2ε, since, when the magnetization increases in a, it decreases in b). We write:

$$\varepsilon=c\frac{J}{J_S} \tag{21}$$

where c is a constant. Substituting these results into Eq. (19) and also using the result (Chikazumi, 1964, p. 211) that $\sum_{m\ \mathrm{odd}}\frac{1}{m^3}\simeq1.0517$, we get

$$E\simeq\frac{4.2 J_S^2 d^3}{\mu_0\pi^3}\left[\frac{1}{d^2}+\frac{2\varepsilon^2}{d^4}+\frac{12\varepsilon^4}{d^6}\right]$$

or

$$E\equiv E_0+\frac{(A+B\varepsilon^2)J^2}{2}=E_0+E_c \tag{22}$$

where E_0, A and B are constants that depend on the domain structure, as given above. In the limit that $\varepsilon = 0$ (this implies $J \rightarrow 0$), $E = E_0 = 4.2 J_s^2 d / \mu_0 \pi^3$, a result obtained by KITTEL (1949) and CHIKAZUMI (1964, p. 211). (The above number is twice that given by these authors, simply because we have obtained the sum of the average energy in *both* domains.) This consistency provides some confidence that no mathematical errors have occurred in the derivation of Eq. (18). From Eqs. (7) and (22) we obtain the demagnetization factor *for small wall displacements*.

$$N = (A + B\varepsilon^2) . \qquad (23)$$

4.   Discussion and Conclusions

Equation (23) can be used to illustrate some important conclusions. First, in the limit of very small domain wall displacements, the demagnetization factor can be taken as a constant, independent of wall displacement. Therefore, the theory leading to Eq. (6) is correct only for small wall displacements. However, small wall displacements are probably to be expected for weak-field TRM. Equally important, the factor N is related to domain structure and not to the shape of the grain. The grain's shape will enter only to the extent that it affects the domain structure.

N increases with further wall displacement. One can calculate what N is close to saturation. (Note, however, that care must be used, if the preceding equations are used to do this. For example, the second and fourth terms in Eq. (17) have factors going to infinity and to zero, as b approaches zero. Care must be used so that the Moore-Smith convergence criterion is not violated.) Complicated mathematical arguments are unnecessary for the limit when saturation is approached, since the grain approaches a single domain configuration. However, contrary to what might be expected at first glance, N is still a function of ε, close to the limit. The limiting value of ε is related to the shape of the domain 'a'. Effectively, what is happening is that the effect of the magnetization in the b domain becomes infinitesimally small as one approaches the limit. Therefore, the change in N close to the limit is well approximated by considering the change in shape of the a domain. In this limit N is the 'shape' demagnetization factor associated with a single domain grain with shape identical to that of domain a.

The above results suggest that the following theorem may be valid: N is a monotonically increasing function of ε. Proof of this theorem is difficult, since it is invalid to extrapolate from particular models of domain geometry to the general situation. Nevertheless, the theorem appears reasonable, even if it remains a difficult one to prove rigorously.

Because N appears to be an increasing function of ε, the acquisition of remanence will not vary simply as a function of the external field raised to some power ($h^{1-1/n}$, as given in Eq. (6)). This prediction is consistent with the log-log plots of J versus h as found by DUNLOP and WADDINGTON (1975) and by DAY (this issue). However, this result does not explain the apparent remarkable consistency between results for alleged single domain TRM and multidomain TRM (DAY, this issue).

In conclusion, the demagnetization factor has been found to depend on domain structure in grains with small wall displacements and on grain shape in grains close to saturation. Previous multidomain theories correctly treat the demagnetization field only in the limit of very small wall displacements. It is difficult to predict the precise way the demagnetization factor will increase in real situations where closure domains, etc., are present. Nevertheless, it seems reasonable to expect that N will generally be a monotonically increasing function of the mean distance of domain wall displacement from the zero field equilibrium configuration.

Critical reviews by Drs. Robert Butler, David Dunlop, Mike Fuller and Frank Stacey greatly improved the quality of this paper. Support for this work came from NSF grants OCE-01541 and OCE 75-21002.

REFERENCES

BUTLER, R.F. and S.K. BANERJEE, Theoretical single-domain grain size range in magnetite and titano-magnetite, *J. Geophys. Res.*, **80**, 4049–4058, 1975.

CHIKAZUMI, S., *Physics of Magnetism*, pp. 554, John Wiley and Sons, New York, 1964.

DAY, R., TRM and its variation with grain size, *J. Geomag. Geoelectr.*, **29**, 233–265, 1977.

DUNLOP, D.J. and E.D. WADDINGTON, The field dependence of thermoremanent magnetization in igneous rocks, *Earth Planet. Sci. Lett.*, **25**, 11–25, 1975.

DUNLOP, D.J. and G.F. WEST, An experimental evaluation of single domain theory, *Rev. Geophys.*, **7**, 709–758, 1969.

KITTEL, C., Physical theory of ferromagnetic domains, *Rev. Mod. Phys.*, **21**, 541–583, 1949.

NÉEL, L., Théorie du traînage magnétique des ferromagnêtiques en grains fins avec applications aux terres cuites, *Ann. Géophys.*, **5**, 99–136, 1949.

NÉEL, L., Some theoretical aspects of rock magnetism, *Adv. Phys.*, **4**, 191–243, 1955.

SCHMIDT, V.A., A multidomain model of thermoremanence, *Earth Planet. Sci. Lett.*, **20**, 440–446, 1973.

SCHMIDT, V., The variation of the blocking temperature in models of thermal remanences (TRM), *Earth Planet. Sci. Lett.*, **29**, 146–154, 1976.

SHIVE, P., The effect of internal stress on the thermoremanence of nickel, *J. Geophys. Res.*, **74**, 3771–3780, 1969.

STACEY, F.D., The physical theory of rock magnetism, *Adv. Phys.*, **12**, 45–133, 1963.

STACEY, F.D. and S.K. BANERJEE, *The Physical Principles of Rock Magnetism*, pp. 195, Elsevier, Amsterdam, 1974.

VERHOOGEN, J., The origin of thermoremanent magnetization, *J. Geophys. Res.*, **64**, 2241–2449, 1959.

Adv. Earth Planet. Sci., **1**, 61–86, 1977

The Hunting of the 'Psark'[*]

David J. DUNLOP

*Geophysics Laboratory, University
of Toronto, Toronto, Canada*

(Received June 17, 1977)

At least four types of sub-domain magnetic moments, on a scale smaller than the main domain structure, could contribute to pseudo-single-domain intensities of TRM (thermoremanent magnetization) in small multidomain grains. Of these, moments pinned by the stress fields of dislocations, surface moments, and moments due to the Barkhausen discreteness of domain wall positions are either strongly shielded by the magnetically soft matrix, subject to the internal demagnetizing field during magnetization changes, or so coupled to the domain structure that they cannot change magnetization independently. Only the net moments of domain walls themselves qualify as 'psarks'—subdomain moments with truly single-domain behaviour. A new reversal mode, domain wall inversion or curling, is postulated to explain the incoherent reversal of domain wall moments. It amounts to nucleating and propagating a Bloch line across a 180° domain wall.

The experimental evidence for the existence of single-domain-like moments within multidomain grains is reviewed. Both initial susceptibility and weak-field TRM have distinctively single-domain components in magnetite grains just above single-domain size. The volume activated in magnetization changes of these same grains, whether determined from TRM acquisition curves, from blocking temperatures or from coercive forces at high temperature, is about equal to the volume of a domain wall, as expected. However, there is no present evidence for psarks in $>1\ \mu$m grains. Psarks are, therefore, a partial explanation at best of the pseudo-single-domain question.

Psarks, if they exist, could have curious behaviour when cooled from above their blocking temperatures, including the possibility of a 'self-rotated TRM' at 90° to the direction of the field applied at high temperature.

1.   Sub-Domain Moments and Psarks

Ferromagnetic domain theory predicts a quantum jump between the magnetic properties of single-domain (SD) and multidomain (MD) grains. Experimentally, no such discontinuity is observed, either in saturation remanence and coercive force (KNELLER and LUBORSKY, 1963) or in TRM (PARRY, 1965; DUNLOP, 1973a). Instead, remanence and coercivity gradually fall, as the grain size increases, from high SD-like values to low, 'true' MD values. In magnetite, at least, the size range over which this pseudo-single-domain (STACEY, 1962, 1963) or PSD behaviour occurs

---

[*]   Paper presented at the special session on the 'Origin of TRM', American Geophysical Union, San Francisco, December 9, 1976.

61

(0.05–15 μm approximately—see Day, 1977, this issue) is much broader than the range for SD behaviour (0.03–0.05 μm in equant grains). The PSD behaviour of magnetite and other ferromagnetics is, therefore, the key to understanding stable TRM (or DRM or CRM, for that matter) in fine-grained rocks.

It seems probable that domain structure in relatively small particles gives rise, intrinsically or incidentally, to permanent magnetic moments on a scale smaller than that of the domains themselves. We are not thinking here of physical subdivision of grains into regions of SD size (Ozima and Ozima, 1965; Strangway et al., 1968; Evans, 1977, this issue) but rather of grains that possess *both* conventional domain structure with MD response to applied fields and some sort of sub-domain structure with SD-like response. Suggested mechanisms for sub-domain moments have included regions in which spins are bound to the stress field of a dislocation (Verhoogen, 1959), residual moments due to Barkhausen discreteness (Stacey, 1963), closure domains and surface terminations of domain walls (Stacey and Banerjee, 1974) and domain walls themselves or wave-like spin arrays resembling the interior of a wall (Dunlop, 1972, 1973a; Dunlop et al., 1974; Stacey and Banerjee, 1974).

Most of these hypothesized sub-domain moments are prey to a common ill: the SD-like response they would display in isolation is compromised by the internal demagnetizing field and the shielding of their MD host. As a result, it is doubtful if they can account for PSD behaviour. To emphasize the fact that sub-domain moments do not, as a rule, have truly SD character, the term 'psark' has been coined (Dunlop, 1976a). A psark is a sub-domain moment that exhibits SD-like behaviour in spite of being immersed in a MD grain. The name is appropriate because psarks, like quarks in elementary particle physics, are unobservable (at any rate with present techniques), their mechanism is problematic and their nature must be deduced from indirect evidence.

2.   Sub-Domain Models

2.1   Spin-pinning by dislocations

Sub-domains of spins deflected from an easy axis of magnetization by the stress fields of dislocations were proposed by Verhoogen (1959). Favourably oriented dislocations produce sub-domains with magnetic moments parallel to the dislocation line, although Shive (1969) showed that pinning by dislocations is insufficient to control stable remanence.

Stacey (1963) pointed out that such moments are substantially shielded by the surrounding domains. Stephenson (1975) has recently reported detailed shielding calculations, but a simpler approach will suffice to make the point. To reduce the magnetostatic self-energy, permanent sub-domain moments will tend to be offset by induced magnetization of the surrounding domains. A sub-domain remanence J_r will give rise to an internal demagnetizing field $-NJ_r$. (Here N is not necessarily the demagnetizing factor of the whole grain and may vary with position in the crystal. For a full discussion, see Merrill, 1977, this issue.) The component of this field,

$-N J_r \cos \phi$, parallel to the domain magnetizations will displace 180° domain walls, with characteristic susceptibility $\chi_\|$, giving an induced magnetization

$$J_{\mathrm{in}} = -N(J_r \cos \phi + J_{\mathrm{in}})\chi_\| , \tag{1}$$

since wall displacement itself modifies the internal demagnetizing field. Solving (1) for J_{in} and adding to it $J_r \cos \phi$, we find for the total magnetization observed parallel to the domains

$$(J_{\mathrm{obs}})_\| = J_r \cos \phi + J_{\mathrm{in}} = J_r \cos \phi (1 + N\chi_\|)^{-1} . \tag{2}$$

Similarly, domain rotation with characteristic susceptibility $\chi_\perp$ partially shields the component of J_r perpendicular to the domain magnetizations:

$$(J_{\mathrm{obs}})_\perp = J_r \sin \phi (1 + N\chi_\perp)^{-1} . \tag{3}$$

The parallel shielding factor, $(1 + N\chi_\|)^{-1}$, is the same as that for MD TRM (DICKSON *et al.*, 1966). Taking N to have its maximum value $4\pi/3$ (larger values of N are, of course, possible, but domains will avoid these directions) and $\chi_\|$ to be about 1 emu cm^{-3} (STACEY and BANERJEE, 1974, p. 73–74), the parallel shielding factor is about 1/5. Since it is much more difficult to rotate domains than to displace 180° domain walls, at least in small fields (see BANERJEE, 1977, this issue for an illustration of this principle), $\chi_\perp$ is normally considerably smaller than $\chi_\|$ and moments (or components of moments) perpendicular to the domains are not as effectively screened as those parallel to the domains.

Shielding reduces the effective intensity of dislocation-line moments by up to 80%, but it does not in itself disqualify the residual moments as possible psarks. The real difficulty is that each dislocation-line moment is exchange coupled to its host domain and cannot behave independently. In particular, it cannot reverse its moment unless the surrounding domain reverses as well.

Thus, although stress-pinning of spins can impart a permanent moment to a MD grain, this moment can reverse only when it is traversed by a 180° domain wall, in low fields anyway. Such moments possess only MD reversal modes, which are subject to the powerful braking effect of the internal demagnetizing field. They cannot be psarks.

2.2   Barkhausen discreteness

Figure 1 illustrates the phenomenon of Barkhausen discreteness (STACEY, 1963). In small grains at zero field, minimum total energy may not correspond to zero magnetization (i.e. minimum magnetostatic self-energy) because variations in the wall energy have a 'wave-length' that is an appreciable fraction of the particle dimensions. The effect is not confined to two-domain particles, but in larger grains with more domains, the average moment of any wall increases as the area of the wall, i.e., as d^2, while the saturation moment of the grains increases as d^3. Added to this intrinsic d^{-1} dependence is the mutual interaction of the Barkhausen moments, which will be such as to minimize the resultant moment.

Even in two-domain grains, Barkhausen moments, like stress-controlled sub-

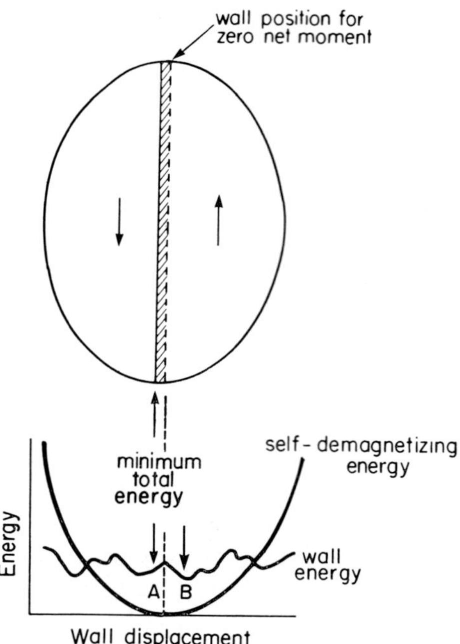

Fig. 1. Barkhausen discreteness of domain wall positions may lead to an imbalance of domain magnetizations (shaded area). Reversal of the Barkhausen moment illustrated requires displacement of the domain wall from A to another minimum B in the total energy.

domains, are largely self-shielded and possess only MD reversal modes. The domain wall will not equilibrate at a position of minimum wall energy. Rather, the internal demagnetizing field will drive the wall up the central barrier in the wall energy almost to the zero magnetization state. Since Barkhausen moments are parallel to domains (they are, after all, merely residual MD moments), the parallel shielding factor $(1 + N\chi_{\parallel})^{-1}$ applies.

To reverse the Barkhausen moment in Fig. 1, the wall must move from A to B. The magnetization passes through zero and is subject to the usual restoring force of the internal demagnetizing field. (Self-shielding is, of course, a manifestation of this restoring force.) Thus Barkhausen moments do not reverse independently. They are coupled to MD processes of the entire grain and cannot be psarks.

2.3 Surface domains and sub-domains

Parts of closure domains and near-surface domain-wall edges may be pinned by surface anisotropy and defects in a manner analogous to the dislocation-pinning of interior sub-domains. (For a full discussion, see BANERJEE, 1977, this issue.) The pinning may be quite strong, as the domain-structure photographs of SOFFEL (1977, this issue) attest. Since surface sub-domains have little freedom of adjustment, they should be less affected by mutual magnetostatic interaction than are Barkhausen moments. Furthermore, since they are generally oriented more or less perpendicular to the interior domains, they are probably poorly shielded as well.

However, it is difficult to imagine how surface moments can reverse independently of MD processes in the interior domains. Closure domains reverse only when an interior 180° wall passes their location, and domain wall edges, if firmly pinned, do not change at all in small displacements of the main body of the wall. (The photographs of SOFFEL, 1977, this issue, leave no doubt that domain walls can bend.) Surface moments, then, seem to possess either MD reversal modes or no reversal modes at all. It is doubtful that they can act as psarks.

2.4 Domain wall moments

DUNLOP (1972, 1973a), arguing from the calculations of AMAR (1958), hypothesized that particles just above SD size are largely filled by a 180° domain wall. Such a wall-like or wave-like spin structure possesses a moment that is a substantial fraction of the SD moment. STACEY and BANERJEE (1974) pointed out that all domain walls, broad or not, possess magnetic moments, oriented perpendicular to the domains. Wall moments and Barkhausen moments have similar average intensities (STACEY and BANERJEE, 1974, p. 61–62) and identical d^{-1} intrinsic grain size dependences of moment per unit volume.

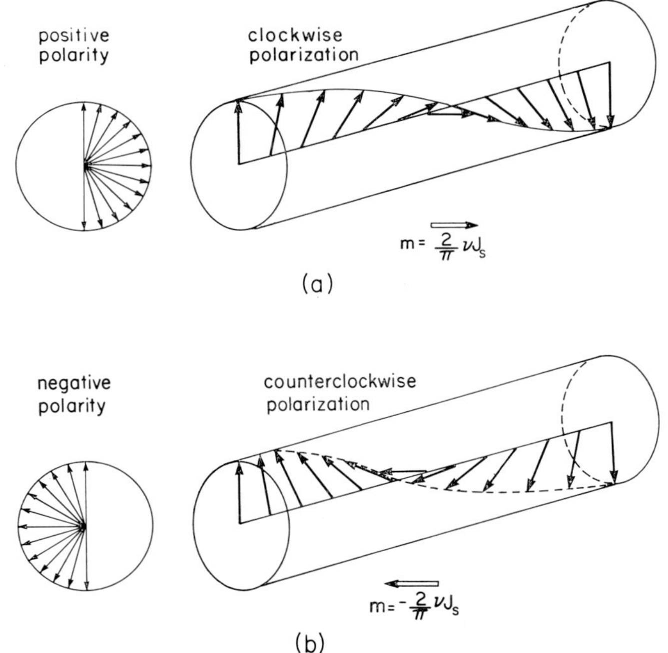

Fig. 2. A single line of exchange-coupled spins along the normal through a 180° Bloch wall. Opposite senses of rotation (describable as clockwise and counterclockwise spin-wave polarizations) lead to net magnetic moments of opposite signs. Both have magnitudes $(2/\pi)vJ_s$, where v is the volume of the wall.

However, there the similarity ends. Figure 2 illustrates that a domain-wall moment has the choice of two polarities. If the spins rotate in a clockwise sense in passing from a domain of positive (upward) magnetization to one of negative magnetization, the net moment is $(2/\pi)vJ_s$ (STACEY and BANERJEE, 1974, p. 61) to the right (positive), v being the volume of the wall and J_s the spontaneous magnetization. Such a wall can also be described as right-handed (SHTRIKMAN and TREVES, 1960), or, since a 180° wall is analogous to a spin-wave of half wavelength, as having a clockwise polarization. Counter-clockwise or left-handed walls between positive and negative domains have net moments that are negative or directed to the left, again of magnitude $(2/\pi)vJ_s$.

Figure 3(a) shows side and end views of a two-domain particle with negative and positive wall moments respectively. Figures 3(b)–3(d) illustrate the three possible responses of a negative wall moment (left-hand figure in each case) to fields applied parallel or perpendicular to the domains. The response to an arbitrarily directed

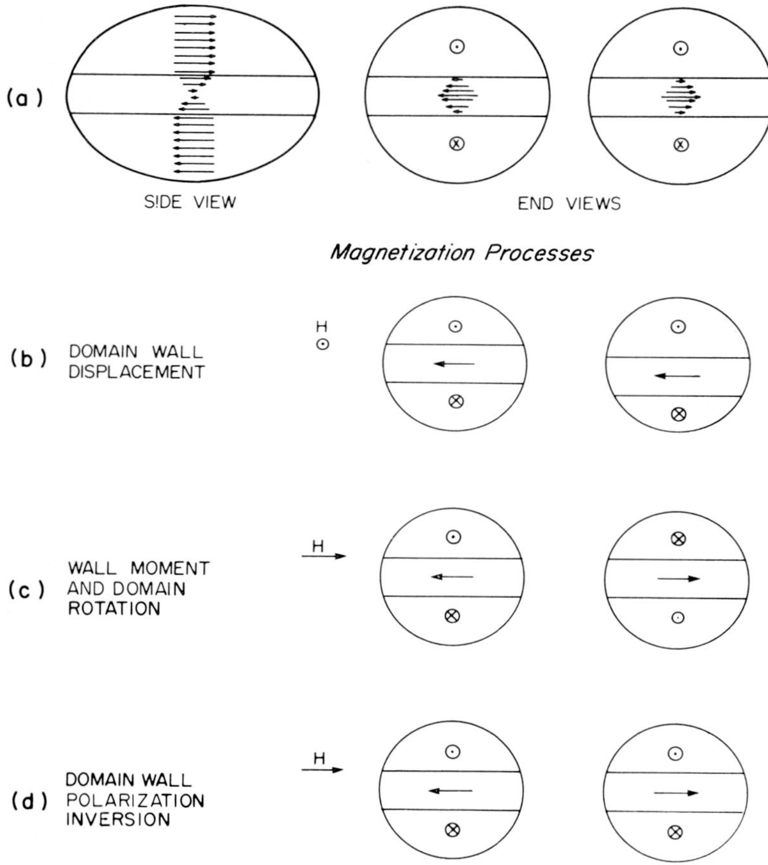

Fig. 3. a) The two possible polarities of a 'psark' or PSD wall moment. b), c), and d) Possible magnetization processes in response to magnetic fields parallel or perpendicular to the domains.

field, with parallel and perpendicular components, will be some combination of these three processes.

In response to a field parallel to one of the domains, the wall is displaced. The wall moment is unchanged in both direction and magnitude. In other words, wall moments coexist with and are unaffected by conventional MD processes. As a special case, wall moments and Barkhausen moments are likewise perpendicular and mutually independent. Wall moments, therefore, have one of the characteristics of a psark: they are unaffected by changes in the magnetization of their multidomain host.

In reality, wall moments are not quite independent of the surrounding domains. Both domain magnetizations will rotate slightly (in the plane of the wall) to produce a moment opposite to the wall moment. However, as we have remarked earlier, perpendicular shielding is relatively ineffective.

It only remains to be seen whether it is possible for the wall magnetization to change in SD fashion without affecting the (nearly) perpendicular magnetization of the multidomain host. Figure 3(c) illustrates wall moment rotation in the plane of the wall. (Exchange coupling precludes rotation in any other plane.) Unfortunately, the domains, being exchange coupled to the wall, also rotate. We have here a truly SD process, which amounts to coherent rotation of the entire volume of an incoherently magnetized particle! Such a reversal mode is, energetically speaking, forbidden. It is well known that a particle may reverse incoherently (by curling or buckling) while retaining a coherently magnetized (SD) remanent state. Therefore, still larger particles, whose remanent state is incoherent will certainly possess an incoherent reversal mode of lower energy than coherent rotation.

This lower energy mode will be called domain wall (polarization) inversion (Fig. 3(d)). The end-point states are certainly stable—they are the same states shown in Fig. 3(a). We will postulate that at some critical value of field H, an instability condition is reached (as with coherent rotation or curling of a SD particle) and one state is more or less instantaneously transformed into the other, reversing the wall moment.

There are two questions to be answered.

1)   Does a wall inversion instability mode exist?

2)   Can the domains and the wall be decoupled during inversion?

As the next section will show, the answer to both questions is yes. A wall moment does have the properties of a psark.

3.   Domain Wall Inversion

3.1   The saturation of a three-domain grain

The following indirect argument demonstrates that domain wall inversion must be possible. Figure 4(a) shows a single line of spins normal to the walls in a three-domain grain. The walls have opposite polarizations (this is actually an energetically unlikely situation because the wall moments add). A near-saturation field has been applied vertically upward. In Fig. 4(b) and 4(c), the field is increased until saturation

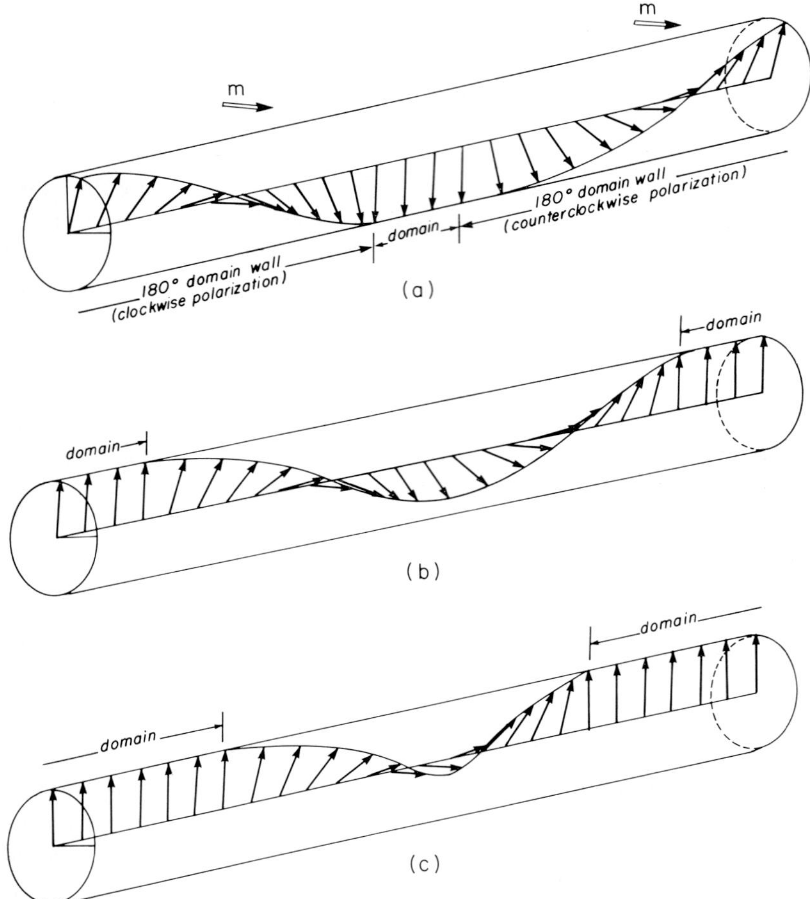

Fig. 4.   The response of a 3-domain grain containing 180° walls of like polarization to a field
applied parallel to two of the domains (vertically up in this diagram). At saturation, the
walls annihilate. A single line of spins along a normal to the walls is shown.

is all but attained. It is clear that the central, unfavourably oriented domain shrinks
and the two oppositely polarized walls coalesce and eventually annihilate.

Figure 5 illustrates the energetically favoured situation: the walls have like polarization and oppositely directed moments. In this case, a saturating field causes the
180° walls to unite into a 360° wall, which cannot be driven out of the grain, since
it has no net moment and the field exerts no pressure on it.

360° walls have not been observed experimentally. This negative evidence is
inconclusive, but other considerations make it unlikely they exist. Suppose that the
field applied to a grain that is saturated, apart from a 360° wall, is reduced to zero
and a small field is applied in the opposite direction. The 360° wall will immediately
and spontaneously subdivide into two 180° walls, renucleating a negative domain as
in Fig. 5(a). If 360° walls existed, ferromagnetics could not be saturated: the domain
structure would be renucleated in a vanishingly small reverse field. It is well known

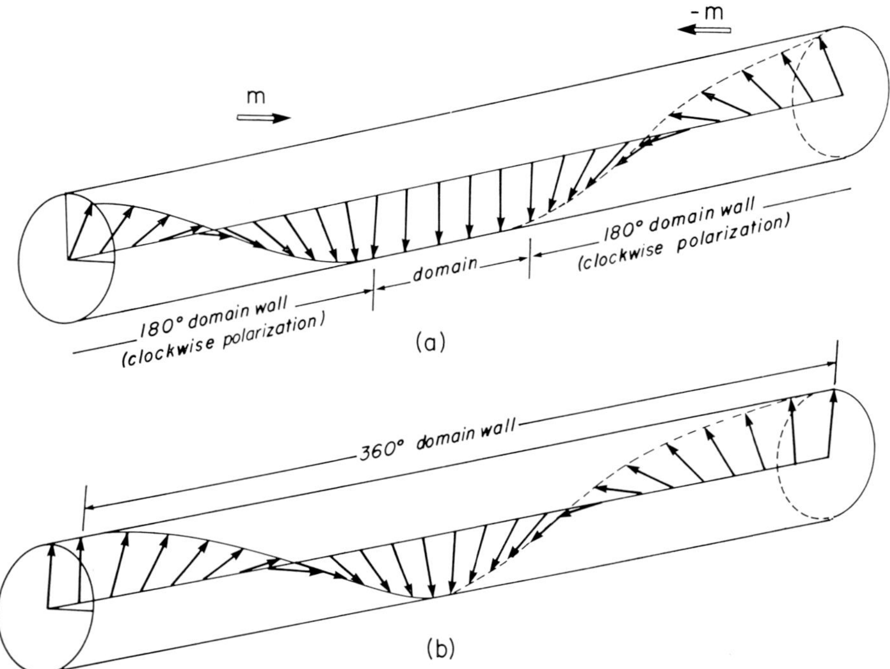

Fig. 5.   180° walls of opposite polarization in a 3-domain grain unite to form a 360° wall at
saturation.

that except in nearly perfect crystals, nucleation fields are less than predicted theoretically, an observation sometimes referred to as Brown's paradox (BROWN, 1945).
Nevertheless, nucleation fields are finite.

The only obvious solution to this quandary is to assert that domain wall inversion must be possible. If either wall in Fig. 5 inverts its polarization, the walls will
annihilate, as in Fig. 4 and the grain can be saturated.

3.2   Wall domains

The actual mechanism by which walls invert can be deduced from the observations of DE BLOIS and GRAHAM (1958), who found that in comparatively large grains
of iron, the 180° (Bloch) walls are subdivided into a periodic sub-domain structure of
clockwise and counter-clockwise wall segments. These 'wall domains' are separated
by 'Bloch lines' whose net moments are inferred to be normal to the magnetizations
of both the wall domains and the main domains. Figure 6 illustrates the structure
observed.

SHTRIKMAN and TREVES (1960) gave a theoretical interpretation, likening the
wall 'domains' to the domain structure in a thin film. Since they rotate in the plane
of the Bloch line, all spins in a Bloch line have a component normal to the domain
magnetizations. Bloch lines, therefore, have high magnetostatic and exchange energies. Minimizing the sum of these energies (the crystalline anisotropy energy of the

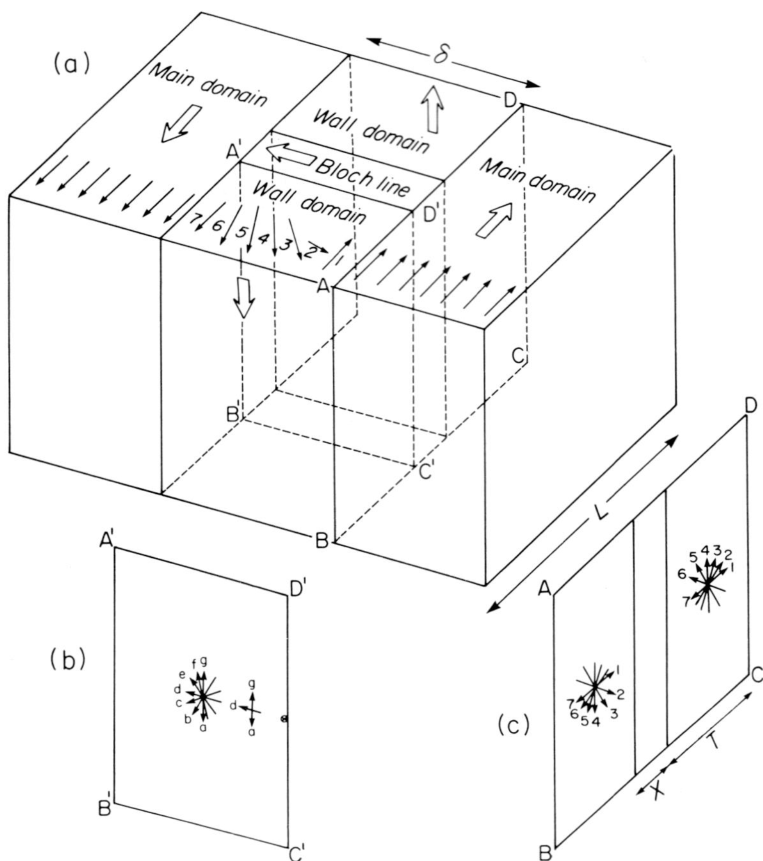

Fig. 6. The subdivision of a 180° wall in a 2-domain grain into two 'wall domains' of opposite polarizations, separated by a 'Bloch line'. The net moments of the main domains, the wall domains and the Bloch lines are indicated by large arrows in (a). The rotation of neighbouring spins in the two wall domains is shown in (a) and (c). The rotation of spins along a normal to the plane of the Bloch line, in passing from one wall domain to the other, is indicated (in projection on the plane A'B'C'D') in (b) for three positions (centre, edge and midway between) within the wall. The spins in a Bloch line rotate about an axis parallel to the magnetization direction of the main domains, i.e., they turn in the plane of the Bloch line. A spin like 4 rotates through 180° (a through g represent its successive positions), a spin like 2 or 3 rotates through less than 180°, and a spin like 1 does not rotate at all in passing through a Bloch line, as shown by comparing (b) and (c).

wall is unaffected by the presence or absence of a Bloch line, and so does not enter this calculation), Shtrikman and Treves find that the Bloch line has an equilibrium width

$$X = [\pi^2 A^{3/2}/8\sqrt{2}\, J_s^2 K^{1/2}]^{1/3} , \tag{4}$$

A and K being the exchange and magnetocrystalline anisotropy constants respectively. The other free dimension of a Bloch line, namely the width of the Bloch wall, is scarcely affected by the introduction of the Bloch line.

Bloch lines occur because the alternation of wall domain polarities greatly reduces the magnetostatic energy of the wall, just as the alternating polarities of ordinary domains reduce the magnetostatic energy of the entire crystal. The number of wall domains (i.e. the ratio L/T—see Fig. 6) is found by minimizing the sum of Bloch line and wall magnetostatic energies (SHTRIKMAN and TREVES, 1960):

$$L/T = (2/\pi)^{1/3}(J_s^2/K)^{2/3} \,. \tag{5}$$

Substituting for magnetite $A = 1.5 \times 10^{-6}$ erg cm^{-1} (GALT, 1952), $K_1 = 1.3 \times 10^5$ erg cm^{-3} (FLETCHER and O'REILLY, 1974) and $J_s = 485$ emu cm^{-3}, we find $X = 265$ Å and $T = 0.78L$. The first figure suggests that Bloch line formation will be difficult in grains just above critical single-domain size ($d = 500–1,000$ Å) where the wall width is ~ 500 Å. The second figure indicates that Bloch lines may be inhibited from forming in magnetite grains of any size, since the favoured number of wall domains is only 1–2 compared 3–4 for iron (SHTRIKMAN and TREVES, 1960), where magnetostatic energy is much more significant due to the larger value of J_s.

The calculations that lead to Eqs. (4) and (5) are approximate and in the case of the magnetostatic energy of the wall, appropriate to large grains. A proper estimate of L/T in small grains would require a calculation of the magnetostatic energy of the wall, subdivided into domains, by the method of AMAR (1958) and BUTLER and BANERJEE (1975). Such a calculation is beyond the scope of this paper.

However, we may note in passing that the number of wall domains observed in iron by DE BLOIS and GRAHAM (1958) was always $\leq$ the 3–4 calculated by Shtrikman and Treves. The reason put forward for this discrepancy was the nucleation energy of a Bloch line. In magnetite, we may presume the number of wall domains is more likely to be 1 than 2. 'Single-domain walls' are, therefore, a very real possibility in small (two-domain or three-domain, say) grains.

3.3 *Inversion of a 'single-domain wall'*

The reversal mode of a two-domain wall is presumably Bloch line displacement. This process calls into play the internal demagnetizing field *of the wall itself* (although not, to a first approximation, that of the main domains) and so, has essentially multi-domain character.

If Bloch walls in small grains were subdivided into domains, the dilemma of how to saturate a three-domain grain (see Figs. 4 and 5) would not arise, since each wall would contain segments of *both* polarizations. Only in a three-domain particle containing single-domain walls is inversion of one of the walls a necessary preliminary to saturating the particle.

The reversal mode of single-domain grains just below critical size for subdivision into two domains is curling (FREI et al., 1957), a process that closely resembles nucleation of a domain wall (whose nucleation energy is unavailable in the absence of an applied field) which more or less instantaneously sweeps across the crystal, reversing the moment of the crystal. Since the 'domain wall' imagined here fills the entire grain (that is all spins rotate, but not in a coherent fashion), the net magnetization of the grain decreases very little during reversal from its saturation (SD) value. The predominant effect is the *rotation* of a moment of slightly reduced magnitude, an essentially SD mode.

I propose that a single-domain wall can invert its polarization (reverse its moment) in an exactly analogous manner, by creation of a Bloch line filling all or most of the wall (the magnetostatic energy of the applied field supplying the necessary nucleation energy), which then sweeps across the wall and vanishes. This hypothesis is speculative but reasonable. The micromagnetic (BROWN, 1963) calculations necessary to demonstrate that such an instability mode exists have not been done and are probably very difficult.

For the present, I will assume that inversion of a single-domain wall is possible and that it occurs by means of 'wall curling' (creation and propagation of a Bloch line across a 180° wall). If so, the moment of a single-domain wall in a two-domain particle satisfies all the requirements for a psark. Furthermore, it is at present the *only* candidate for a psark.

4. Characteristics of 'Psarks'

4.1 *PSD and MD TRM*

Any permanent sub-domain moment can contribute to the PSD enhancement of weak-field TRM, but only 'psarks', which reverse their moments in SD-like fashion, can obey STACEY's (1963) PSD theory of TRM. Stacey proposes PSD moments with only two possible states, oriented at angles θ or $\pi-\theta$ to an applied field $\boldsymbol{H}$. The average component of TRM parallel to $\boldsymbol{H}$ is then (STACEY, 1963, Eq. 5.15)

$$J_{\mathrm{trm}} = \frac{m}{v} \cos\theta \, \tanh\left[\frac{m(T_B)H}{kT_B}\cos\theta\right],\tag{6}$$

m and v being the PSD moment and volume respectively, and T_B the blocking temperature. Averaging over m and θ (STACEY and BANERJEE, 1974) does not much alter the aspect of the TRM acquisition curve.

Equation (6) is a strictly SD formulation, written down directly from the NÉEL (1949) SD theory. Such a formulation implies no intermediate states with moments between $\pm m$, whereas any MD magnetization possesses such states. Nor is a parasitic sub-domain moment that cannot be decoupled from the main domains permitted. The PSD moment must reverse freely (above a critical field or temperature) between its equilibrium states. Otherwise a potential energy of coupling to the host domains would enter the Boltzmann factors which are the essence of Eq. (6). Only psarks have the required SD-like characteristics.

The main domains exhibit a MD TRM given by (STACEY, 1963, Eq. 5.6)

$$J_{\mathrm{trm}} = \frac{H}{N}\frac{J_{SO}}{J_S(T_B)}\frac{1}{1+N\chi_{\|}}$$

or, eliminating the implicit dependence of $J_{SO}/J_S(T_B)$ on H (NÉEL, 1955; SCHMIDT, 1973; DUNLOP and WADDINGTON, 1975), by

$$J_{\mathrm{trm}} = C \frac{H^{1/n} H_{CO}^{1-1/n}}{N} \frac{1}{1+N\chi_{\|}} . \tag{7}$$

In Eq. (7), C is a number in the range 1–2, H_C is coercive force (subscript O indicates room temperature) and n is the index in the approximate relation $H_C(T) \propto J_S^n(T)$. The factor $1/N$ reminds us of the role played by the internal demagnetizing field in MD TRM, as does the shielding factor $(1+N\chi_{\|})^{-1}$. Strictly speaking, an analogous but smaller shielding factor $(1+N\chi_{\perp})^{-1}$, should appear in Eq. (6).

Equations (6) and (7) behave quite differently. PSD TRM is intense at quite low fields and saturates rapidly. MD TRM predominates at high fields. Analysing TRM's produced in a variety of fields is, therefore, a possible means of resolving psarks against their MD background.

4.2   PSD and MD susceptibility

MD grains have an initial susceptibility (STACEY, 1963, Eq. 4.14)

$$\chi_0 = \frac{1}{3} \frac{\chi_{\|}}{1+N\chi_{\|}} + \frac{2}{3} \frac{\chi_{\perp}}{1+N_{\perp}} . \tag{8}$$

(Note that the shielding factors are the same as discussed earlier.) The internal demagnetizing field limits χ_0 to values $\leq 1/N$ (≤ 0.25). Equation (8) shows that this maximum value, which is independent of grain size and coercive force, is achieved when $\chi_{\|}$ and $\chi_{\perp}$ are both $\gg 1/N$. $\chi_{\|}$ is about 1 and satisfies this condition reasonably well. $\chi_{\perp}$ in 2- or 3-domain grains may be as small as 0.25 (STACEY, 1963, Eq. 4.21) and does not satisfy the condition. Thus, the MD susceptibility of small grains will be somewhat less than $1/N$ and may have a slight grain size or coercive force dependence.

Wall moments have SD-like initial susceptibility by virtue of reversible rotations in the plane of the wall (Fig. 3c). Domain-wall inversion contributes nothing to the induced magnetization unless the critical field for nucleation of a Bloch line is exceeded.

The initial susceptibility of SD grains, each with volume v, moment $m = vJ_S$, and microscopic coercive force or critical field H_K for coherent rotation, is (NÉEL, 1949)

$$\chi_0 = \frac{2}{3} \frac{m}{vH_K} = 0.349 \frac{J_S}{H_R} . \tag{9}$$

The final expression makes use of the fact that remanent coercive force H_R for randomly oriented SD moments is $0.524 H_K$ (STONER and WOHLFARTH, 1948).

Equation (9) cannot be carried over directly to the case of psarks for a number of reasons. First, $m = (2/\pi) v_{\mathrm{wall}} J_S$ for psarks, so that χ_0 is reduced by a factor $(2/\pi)$ $(v_{\mathrm{wall}}/v_{\mathrm{grain}})$. Second, wall moments are restricted to a plane in their rotations and so will have a smaller χ_0 than SD moments with unrestricted rotations. (An analogous situation is the basal plane ferromagnetism of hematite—see WOHLFARTH, 1955 or STACEY, 1963, Eq. 4.30). Third, for psarks, H_K for coherent rotation is appropriate for susceptibility whereas H_R is determined by the unknown, but certainly smaller, critical field for wall inversion. The relationship between χ_0 and H_R is further obscured by the fact that H_R as observed experimentally is an average, with unknown weighting of MD and PSD contributions.

Despite these complications, psarks should have an initial susceptibility that is more clearly dependent on coercive force (and hence on grain size) than is MD susceptibility, although a precise $(H_R)^{-1}$ dependence is not be expected. This PSD susceptibility is added to and independent of the MD susceptibility. It provides another way of detecting psarks immersed in MD surroundings.

4.3 *Thermal activation of SD, PSD and MD moments*

Initial susceptibility is largely reversible, but most processes of interest in rock magnetism entail irreversible changes in magnetic state. These changes may be either field controlled or thermally activated. (In practice, this distinction is blurred: even room-temperature isothermal remanence or IRM in quite high fields is thermally aided to a significant extent in small MD grains.) Thermally activated processes, like TRM, VRM (viscous remanence) and IRM in small grains, have well-defined relaxation times, all of which depend strongly on the thermal excitation or *activation volume* within which spins rotate during the magnetization change and the threshold or *critical field* that would characterize the change in the absence of thermal agitation. The activation volumes and critical fields of SD, PSD and MD moments are distinctly different and provide another potential means of detecting psarks.

SD grains have a relaxation time for thermally activated coherent rotation or curling reversals (NÉEL, 1949)

$$\frac{1}{\tau} = \beta f_0 \exp\left[-\frac{v J_S H_K}{2kT}\left(1 - \frac{|H|}{H_K}\right)^2 \right] \tag{10}$$

where $f_0 \approx 10^9$ sec^{-1} (MCNAB *et al.*, 1968). $\beta = 2$ if $H = 0$, $\beta = 1$ if $H \gtrsim 3kT/vJ_S$ and $1 < \beta < 2$ if $0 < H \lesssim 3kT/vJ_S$.

In MD grains, for small wall displacements near a state of zero net magnetization, the internal demagnetizing field can be neglected. The relaxation time, using NÉEL's (1950, 1955) dispersion field model, is then

$$\frac{1}{\tau} \sim f_0' \exp\left[\frac{3A\lambda J_S H_C}{8\pi kT}\left(1 - \frac{|H|}{H_C}\right)^2 \right], \tag{11}$$

in which $f_0' \approx 10^{10}$ sec^{-1} (NÉEL, 1955, p. 237), H_C is the coercivity due to barriers to wall displacement, and $A\lambda$, the product of wall area and the length of one Barkhausen jump, is the appropriate activation volume for the MD process. The STREET and WOOLLEY (1949) barrier fluctuation model leads to a similar expression involving $(H_C - H)$ instead of $(H_C - H)^2$. The difference is not crucial since neither model applies unless H is very nearly zero. Relaxation times in larger fields, where internal demagnetization is important, are considered by STACEY (1963, p. 77–79) and DUNLOP (1973b, p. 873–879).

For psarks, the relaxation time for wall inversion can be written down by analogy with the SD case:

Table 1.   Activation volumes v_{act} and critical fields H_K (or H_C) in relaxation time expressions of the form $1/\tau = \beta f_0 \exp\left[-\alpha \dfrac{v_{act} J_S H_K}{kT}\left(1-\dfrac{|H|}{H_K}\right)^2\right]$ (α is a number of order 0.2–1).

	SD (Eq. 10)	Psark (Eq. 12)	MD (Eq. 11)
v_{act}	$v_{grain} = v$	$v_{wall} = A\delta$	$v_{Barkhausen} = A\lambda$
$H_K : H_C$:	coherent or incoherent rotation	wall inversion	wall displacement

$$\frac{1}{\tau} = \beta f_0'' \exp\left[\frac{(2/\pi) A\delta J_S H_K}{2kT}\left(1-\frac{|H|}{H_K}\right)^2\right]. \qquad (12)$$

Likely $f_0'' \approx f_0$, since both are inverse atomic correlation times for similar thermal excitations. The wall volume, $A\delta$ (see Fig. 6), is the appropriate activation volume and H_K is the nucleation field for wall inversion.

The significant point is that coherent or incoherent rotations of an SD grain, MD wall displacements, and PSD wall inversions have relaxation times with identical dependences on T, H, v and H_K (or H_C). Data for thermally activated magnetization changes can be analysed in the same manner for all domain structures. The values of v and H_K that are deduced are diagnostic of the domain structure. Table 1 summarizes the activation volumes and critical fields that appear in Eqs. (10) through (12).

5.   Experimental Evidence for Psarks

5.1   *Initial susceptibility*

Figure 7 compares initial susceptibility data for four samples containing sized,

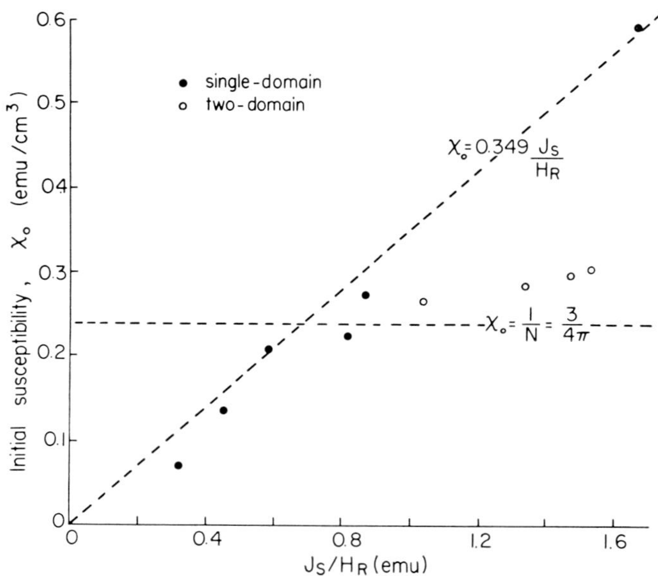

Fig. 7.   Observed initial susceptibilities of single-domain and 2-domain iron oxide grains compared with SD and MD predictions (Eqs. 8 and 9 of the text). (After DUNLOP, 1977).

　　　　　　　　　　　　　D.J. DUNLOP

slightly cation-deficient magnetite crystals with similar data for six samples containing SD iron oxides of various types. The former samples contain roughly cubic crystals with mean sizes of 370, 760, 1,000 and 2,200 Å respectively. (See DUNLOP, 1973a, c for a full description of the preparation and properties of these magnetites.) Except in the finest-grained sample, all grains are believed to be above SD size and probably 2-domain (DUNLOP, 1972, 1973a). The SD samples are described by DUNLOP and WEST (1969).

The SD samples follow Eq. (9) quite closely. The 2-domain samples clearly do not. However, their susceptibilities *do* depend on grain size and all are larger than $1/N$. These are the characteristics of a psark contribution to susceptibility in a small MD grain.

5.2　TRM acquisition curves

Figure 8 shows that TRM vs. H curves measured for the same four magnetite samples have a dependence on grain size that is relatively weak but in the direction predicted for PSD TRM by Eq. (6). MD TRM, according to Eq. (7), increases with increasing H_C (i.e. decreasing grain size), the opposite trend to that observed. The theoretical PSD type curves, calculated from a refined version of Eq. (6), have a pronounced size dependence, but the comparison is not a fair one because the model

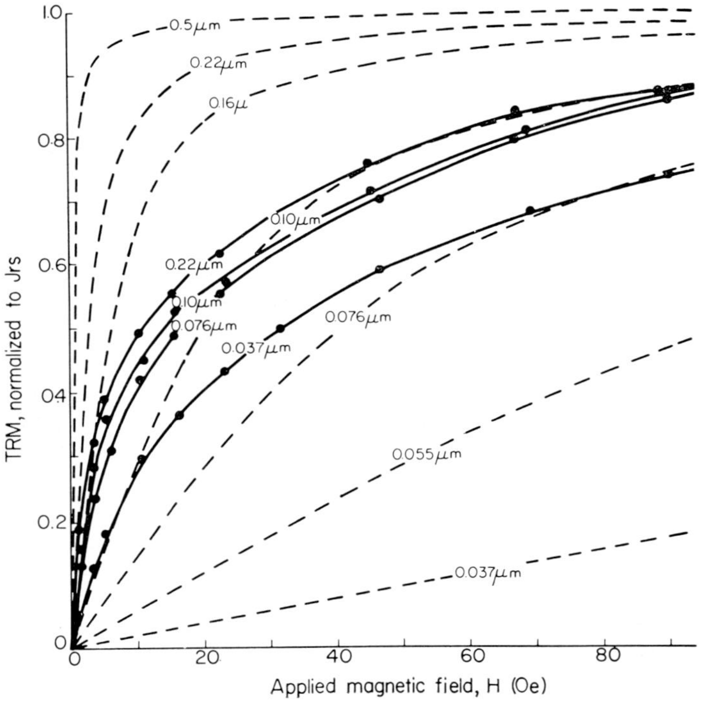

Fig. 8.　Observed field dependences of TRM in four magnetite samples with mean grain sizes 370, 760, 1,000 and 2,200 Å. The theoretical PSD type curves (dashed) follow Eq. (6) of the text. (Data from DUNLOP, 1973a).

calculations assumed PSD moments due to walls filling entire grains. (This turns out to be a reasonable assumption for the 760 and 1,000 Å grains but not for the 2,200 Å grains.) Wall moments are actually more likely to have a d^2 than a d^3 dependence.

DUNLOP (1973a) and DUNLOP *et al.* (1974) showed that the data of Fig. 8 can be fitted quite closely by a purely MD function (Eq. 7) for fields above 20 Oe, and a sum of MD and PSD contributions below 20 Oe. Because MD TRM is so strongly limited by the internal demagnetizing field, only the psark contribution is significant in fields of ~ 0.5 Oe.

5.3   *Activation volumes*

The curve fits of DUNLOP *et al.* (1974) yielded values of $m(T_B)$ (Eq. 6). Since T_B values were known from other measurements (DUNLOP, 1973a), the activation volumes shown in Fig. 9(a) could be calculated.

In the largest-grained sample, the cube edge, d_{act}, corresponding to v_{act} is 1/2 the observed median grain size $\langle d \rangle$. Assuming 2-domain structure, the wall volume should, therefore, be 1/8 the volume of the grain, on average.

We can test whether or not the wall moment model is reasonable in the following way. The saturation remanent moment of a grain containing one wall psark is $(2/\pi)v_{wall}J_S$. Since grains, and psarks, are randomly oriented, the global saturation moment observed in any direction is 1/2 this figure. The saturation induced moment is $v_{grain}J_S$. Thus

$$J_{rs}/J_S = (1/\pi)(v_{wall}/v_{grain}) \, . \tag{13}$$

For the $\langle d \rangle = 2,200$ Å sample, J_{rs}/J_S should be about $(8\pi)^{-1} \cong 0.04$. Actually, for a field of 1,500 Oe, J_r/J is about 0.1 (DUNLOP, 1972). In truly saturating fields, the value would be somewhat smaller, but there remains a discrepancy of about a factor of 2. Possibly these are 3-domain grains in which each of the two walls can be activated independently.

The $\langle d \rangle = 760$ and 1,000 Å samples have d_{act} values quite close to $\langle d \rangle$. (In the finest-grained sample, d_{act} would also be close to $\langle d \rangle$ if the superparamagnetic fraction below $d = 300$ Å were eliminated before averaging.) However, J_r/J in 1,500 Oe is only 0.2–0.3, about half the SD value. DUNLOP's (1972, 1973a) interpretation is that these samples contain mostly grains just above SD size which are non-uniformly magnetized in zero field but have no recognizable domains. One can imagine the onset of such a state in a large SD grain, as spins at opposite edges of the grain curl in opposite directions relative to the uniformly magnetized core (e.g., see BROWN, 1958), or in a 2-domain grain as the domain wall expands to fill the grain (AMAR, 1958).

We can use Eq. (13) to test whether this model is reasonable. For the $\langle d \rangle = 370$ and 760 Å samples, $v_{wall} \approx v_{grain}$ and J_{rs}/J_S should be $1/\pi \cong 0.32$. The observed J_r/J values are 0.29 and 0.24 respectively. The $\langle d \rangle = 1,000$ Å sample has $d_{act} = 0.82 \langle d \rangle$ and so J_{rs}/J_S should be $0.55/\pi \cong 0.18$. $J_r/J = 0.20$ is observed.

Considering how rough are the determinations of d_{act}, even this approximate agreement with model values for psarks is encouraging.

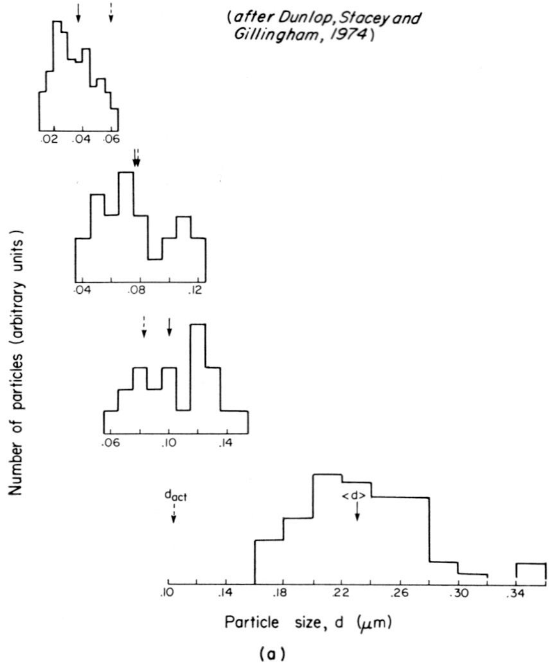

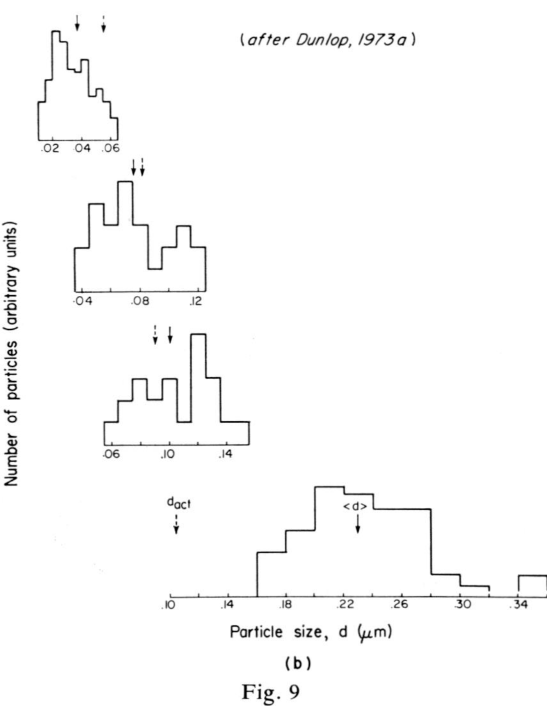

Fig. 9

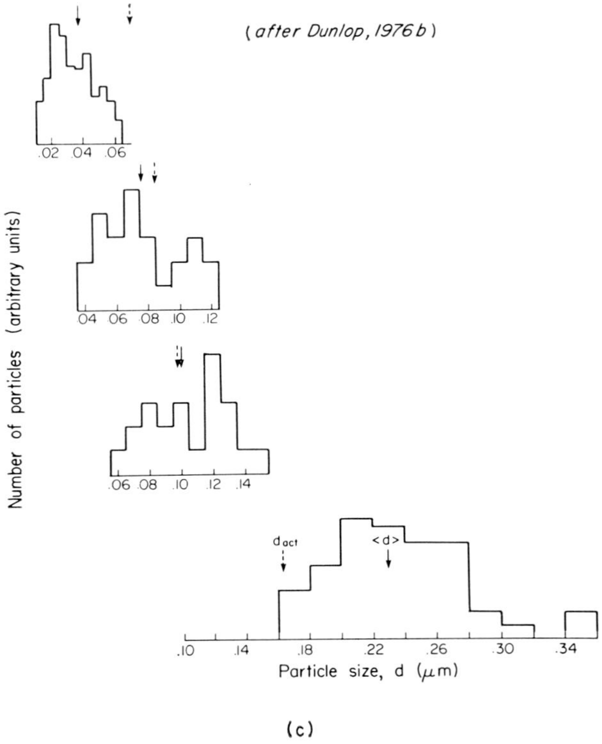

Fig. 9.   Grain-size distributions observed by electron microscopy (median values shown by solid arrows) for the magnetite powders of Fig. 8, compared with thermal activation volumes (equivalent cube edge shown by dashed arrows) calculated from: (a) analysis of the field dependence of TRM intensity (DUN-LOP *et al.*, 1974); (b) observed blocking temperatures of TRM (DUNLOP, 1973a); and (c) thermal fluctuation analysis of high-temperature coercive force data (DUNLOP, 1976b).

An alternative hypothesis is that the SD threshold is $>1,000$ Å and that values of $J_{rs}/J_S<0.5$ result from particle interaction (DAVIS and EVANS, 1976). If this were the case, however, one would expect J_{rs}/J_S to be constant or to *increase* slightly with increasing grain size, since the finer the particle size, the smaller the average centre-to-centre particle separation in a cluster. In reality, J_{rs}/J_S *decreases* steadily with increasing grain size, a characteristically MD trend.

The activation sizes shown in Fig. 9(b) are very similar to those of Fig. 9(a) but were determined quite independently from measured blocking temperature data (DUNLOP, 1973a). Since at the blocking temperature T_B, $\tau \approx t$, a typical experimental time, substitution in Eq. (12) gives

$$T_B = \frac{v_{\mathrm{act}} J_S(T_B) H_K(T_B)}{\pi k \, ln(\beta f_0'' t)} \left[1 - \frac{|H|}{H_K(T_B)} \right]^2 \tag{14}$$

where $v_{\mathrm{act}} = A\delta$. Similar equations follow from Eqs. (10) and (11) for SD and MD blocking temperatures. DUNLOP (1973a) determined average blocking temperatures by thermally demagnetizing weak-field TRM's and assumed $H_K = H_R$, the remanent coercive force, in arriving at the d_{act} values of Fig. 9(b).

5.4 Thermal fluctuation analysis

The activation sizes given in Fig. 9(c) were arrived at by a still different method, thermal fluctuation analysis (DUNLOP, 1976b). Isothermal magnetization changes, except at absolute zero, are never wholly field induced. The contribution of thermal agitation makes it fair to view the coercive force H_C as the 'blocking' field at which $\tau \approx t$ for an isothermal (not necessarily room-temperature) magnetization change. Equation (12) then transforms into

$$H_C = H_K - \sqrt{\frac{\pi k T \ln(\beta f_0'' t) H_K}{v_{act} J_S}} = H_K - H_q \qquad (15)$$

for psarks. (Analogous equations for SD grains (BEAN and LIVINGSTON, 1959) and MD grains (NÉEL, 1950, 1955) follow from Eqs. (10) and (11).) H_q is the celebrated NÉEL (1955) fluctuation field.

We lack a fundamental knowledge of how the threshold field for domain wall inversion depends on temperature. Pragmatically, $H_K(T) \propto J_S(T)$ (as for shape-controlled SD grains) fits the H_C data at low temperatures, where H_K outweighs H_q. With this assumption, we can recast Eq. (15) in the useful form

$$\frac{H_C}{j_s} = H_{KO} - \sqrt{\frac{\pi k \ln(\beta f_0'' t) H_{KO}}{v_{act} J_{SO}}} \left(\frac{T}{j_s}\right)^{1/2}. \qquad (16)$$

Here $j_s \equiv J_S(T)/J_{SO}$ and subscript O means 20°C value. Thermal fluctuation analysis in its simplest form is the determination of H_{KO} and v_{act} from the intercept and slope of an H_C/j_s versus $(T/j_s)^{1/2}$ plot.

DUNLOP (1976b) analysed H_C and H_R data from hysteresis loops at temperatures between 20° and 550°C, in arriving at the d_{act} values of Fig. 9(c). These values are all larger than those of Fig. 9(a) and 9(b), especially in the case of the $\langle d \rangle = 2,200$ Å sample. This is hardly surprising because the data for the earlier analyses were taken in fields of 0–20 Oe, while the coercive forces analysed for Fig. 9(c) ranged from 20 Oe at 550°C to a maximum of 395 Oe at 20°C. Fields of this strength may well modify the domain structure of a Bloch wall. Furthermore, MD wall displacement must contribute to the overall coercive force.

We can assess the latter effect, by applying fluctuation analysis to soft (i.e., assumed MD) and hard (psark) fractions of the coercive force spectrum. DUNLOP and BINA (1977) analysed high-temperature af demagnetization curves in this manner. Table 2 gives d_{act} and H_{KO} values obtained from separate analyses of five spectral fractions for the $\langle d \rangle = 760$ Å sample. The fractions had very different median coercivities but essentially similar activation volumes.

In this sample and the $\langle d \rangle = 1,000$ Å sample, whose results are similar, MD wall displacement seems to play no role in coercive force, at least not one that is resolvable from harder, presumably SD-like processes. In retrospect, we might have antici-

Table 2. Comparison of d_{act} and H_K values predicted by thermal fluctuation analysis with experimental data for five coercivity spectral fractions of a PSD magnetite sample ($\langle d \rangle =$ 760 Å).

	Fraction of coercivity spectrum					Mean
	0–20%	20–40%	40–60%	60–80%	80–100%	
Theoretical[1] d_{act} (Å)	780	705	735	755	675	730
Experimental[2] $\langle d \rangle$ (Å)						760
Theoretical[1] H_{KO} (Oe)	197	307	359	420	593	348
Experimental[3] H_{CO} (Oe)	168	250	299	349	473	299[4]

[1] From thermal fluctuation analysis (see text) (DUNLOP and BINA, 1977).
[2] From electron micrographs (DUNLOP, 1973c).
[3] Median af demagnetizing field of each fraction (DUNLOP and BINA, 1977).
[4] Median demagnetizing field of the whole sample, not the mean of the listed H_{CO} values.

pated that in a spin structure that fills all or most of a grain, incoherent rotation (i.e., inversion) and wall displacement would be indistinguishable.

In the $\langle d \rangle = 2,200$ Å sample, where discrete domains have been inferred, the lowest coercivity fraction behaves in a distinctly different fashion from other fractions and fluctuation analysis is not very successful. Here MD and PSD processes seem to be mixed, except at low coercive forces (≤ 100 Oe).

Table 2 shows that wall inversion has a spectrum of critical fields H_{KO} intermediate between the soft (< 100 Oe) values that characterize > 1 μm magnetite and very hard (coherent rotation) SD values. (Af coercivities provide reasonable estimates of these H_{KO} values in all but the finest grains.) The observation of coercivities that seemed implausibly high for MD magnetite of course provided the initial impetus for PSD models (e.g., VERHOOGEN, 1959). On the other hand, Dunlop and Bina's results imply that in ≈ 0.25 μm magnetite, wall motion may be characterized by coercivities as high as 200–300 Oe. If so, the simple-minded distinction between 'soft' MD processes and 'hard' PSD processes does not hold in fine grains.

5.5   Psarks in > 1 μm magnetite grains

DUNLOP and BINA (1977) analysed high-temperature hysteresis and af demagnetization data of a sample containing well-sized and strain-free magnetite grains in the 1–5 μm range. Fluctuation analysis failed in both cases. Instead, $H_c(T)$ was fit very well by a MD power-law expression $\propto J_S^n(T)$, with no fluctuation-field term. Whatever the nature of sub-domain moments in > 1 μm magnetite grains—and to judge by the size dependence of TRM, they apparently exist in the 1–5 μm range (see e.g., DAY, 1977, this issue)—they must be either coupled to the main domains (i.e., not SD-like) or if independent, too large to respond noticeably to thermal agitation.

6.   The Possible Role of Psarks in TRM Blocking

Figures 10(a) and 10(b) are speculative models that trace two possible ways TRM

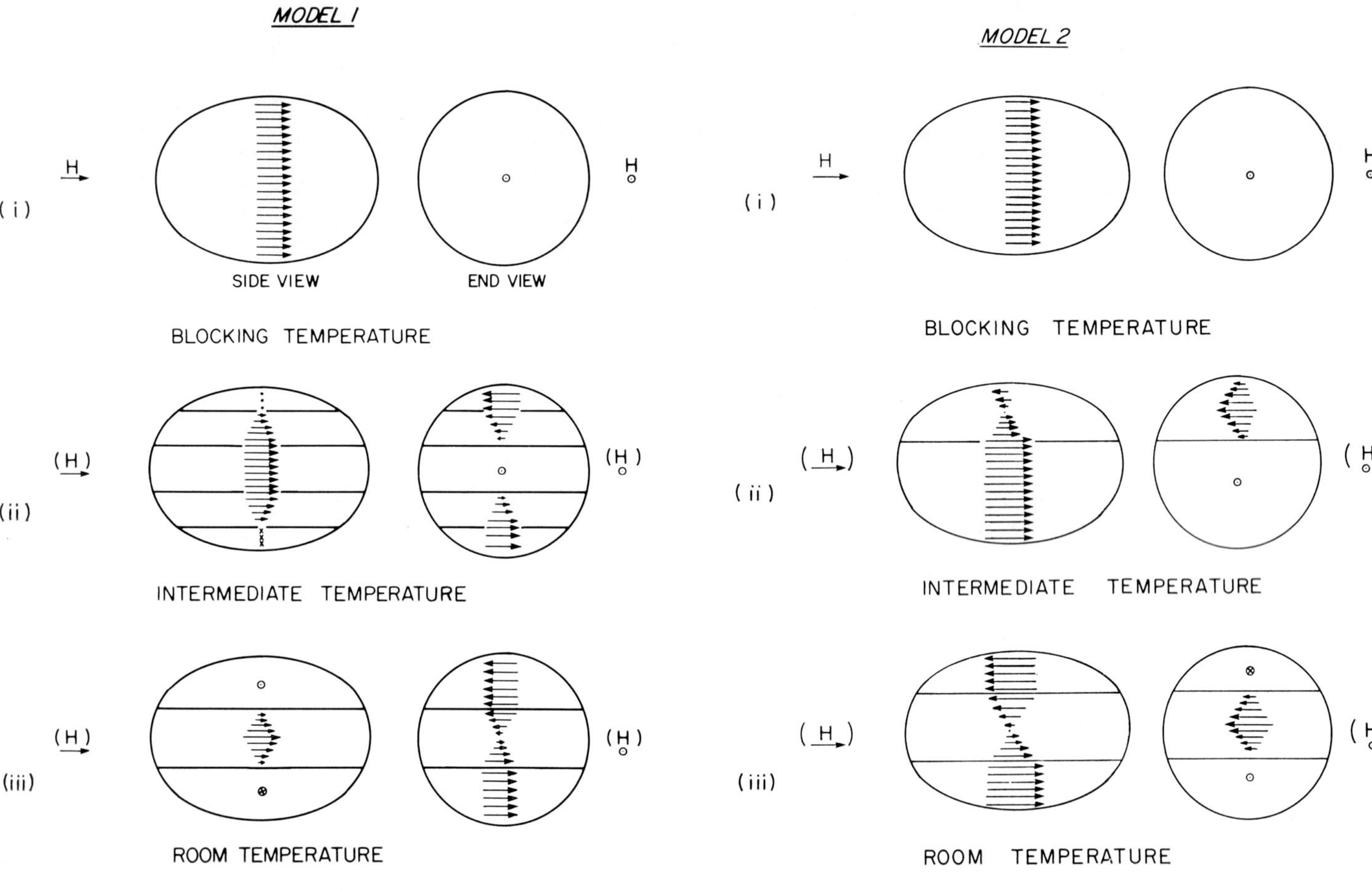

Fig. 10. Two models for the possible development of psark TRM in a 2-domain particle cooled from an SD configuration at the blocking temperature. In model 1, the TRM parallels the field that acted at the blocking temperature. In model 2, the TRM is 'self-rotated' through 90°.

could develop in a 2-domain grain whose room-temperature remanence is predominantly due to the domain-wall moment. BUTLER and BANERJEE (1975) predict that the SD threshold in magnetite moves to larger sizes at high temperature, so that such a grain could be SD near the Curie temperature. (On the other hand, DUNLOP, 1977, reports that for all four magnetite samples studied earlier in this paper, J_r/J *decreases* with increasing T. That is, the domain structure is *less* SD-like at high temperatures than at room temperature.) In any case, both anisotropy and self-demagnetization are so weak near the Curie point that a field $H \sim 1$ Oe should serve to saturate a transitional 2-domain grain above and at its blocking temperature, whether its remanent state is SD or not. This is the initial state pictured in Fig. 10(a)(i) and 10(b)(i).

Below T_B, the presence or absence of a weak field is immaterial. What matters is how the domain structure nucleates. In Fig. 10(a), the main domains, separated from the uniformly magnetized core by 90° walls, nucleate at either edge of the grain. With growth of the main domains, the 90° walls move inwards and eventually coalesce into a 180° wall. (Note that the 90° walls have no choice of polarization and cannot annihilate.) In this model, the psark TRM at room temperature 'remembers' the direction of H at T_B.

In Fig. 10(b), a 180° wall nucleates at one edge of the grain. In the absence of an applied field H, the wall is immediately driven by the internal demagnetizing field to the centre of the grain. One of the domains remembers the uniform magnetization, aligned with H, that existed at and above T_B. The psark, however, is coupled perpendicular to this direction, giving rise to a TRM *perpendicular* to H!

Self-reversed TRM is rare and 'self-rotated TRM', as we might term the above phenomenon, has never been observed. Indeed, since in the model of Fig. 10(b) either domain is equally likely to 'remember' the high-temperature magnetization and a domain wall of either polarity is equally likely to nucleate, there is no reason to expect any net 'self-rotated TRM' below T_B. A possible way of detecting this phenomenon, if it exists, is by applying ever-increasing fields during cooling, until one reaches the threshold field necessary to rotate the spin structure through 90° and align the psark moment with H. Such a field is somewhat analogous to the spin-flopping field of an antiferromagnet.

We shall need to expand greatly our present knowledge of how domain structure nucleates and develops during cooling (see SOFFEL, 1977, this issue) before we can judge whether models like those of Fig. 10 are at all realistic.

7. Summary

Enhanced intensity of TRM in small MD grains could be due to sub-domain moments of many types. Most sub-domain moments, however, are not SD-like. They are either so coupled to the main domains that they cannot reverse unless traversed by a domain wall, or possess no independent reversal mode not limited by the internal demagnetizing field. In addition, most are strongly shielded by the MD matrix.

Only the intrinsic moments of domain walls have SD-like ('psark') character and obey Stacey's PSD theory of TRM. The net wall moment *is* exchange-coupled to the adjacent domains, but because the coupling is *perpendicular*, the wall moment is poorly shielded, exerts no influence on MD magnetization changes (i.e., wall displacement) and in turn is minimally influenced in its *own* magnetization changes by the internal field of the main domains. The final ingredient, a truly SD-like reversal mode, is provided by a postulated instability mode called wall inversion, resembling curling of an SD grain. Being a critical-field reversal, there are no stable intermediate states and the internal demagnetizing field of the wall itself is not brought into play.

Psarks can be detected experimentally, in spite of being superimposed on normal MD magnetizations, in three general ways:

1) by isolating part of the initial susceptibility or weak-field TRM that is not controlled by the internal demagnetizing field (cf. Eqs. 6 through 9);

2) by determining the volume activated in TRM or high-temperature isothermal magnetization and relating it to the volume of one wall (cf. Eqs. 10 through 16);

3) by determining the critical field H_K for magnetization change, by thermal fluctuation analysis or more approximately from af demagnetization, and relating it to the as yet unknown threshold field for wall inversion.

The first two lines of inquiry support the existence of 'psarks' in magnetite grains between 0.05 μm (the SD threshold size) and about 0.25 μm in size. The third criterion is ambiguous because in these fine grain, MD wall displacements may have intermediate (100–300 Oe) coercivities that overlap the range of psark coercivities.

Depending on the manner in which domains nucleate when a 2-domain grain just above SD size (room-temperature state) is cooled from a saturated (SD) state at its blocking temperature, psark TRM may reproduce the field direction or may conceivably be 'self-rotated' 90° from the field direction. Over statistically large numbers of grains, no net 'self-rotated TRM' will be observable below the blocking temperature.

Magnetite grains 1–5 μm in size show no evidence of psarks, yet they exhibit TRM of enhanced 'PSD' intensity. This observation points up a continuing problem that psarks fail to solve. The PSD concept was fostered by the desire to explain theoretically the smooth transition, over a broad grain size span, from SD to MD behaviour. Psarks, on theoretical grounds, are unlikely to be influential except in 2-domain particles because energy minimization will favour psarks with mutually balancing moments in 3-domain and larger grains. The experimental evidence suggests that psark and MD moments begin to merge, so far as coercive force is concerned, in ≈ 0.25 μm magnetite grains.

The quantum step in magnetic properties anticipated above the SD threshold has, therefore, not been averted by the introduction of psarks. It has merely been moved, in modified form, to the 2-domain to 3-domain transition size. This size is not well known, but clearly is well below 15 μm in magnetite, leaving the TRM properties of larger PSD grains unaccounted for.

My work on PSD moments has been aided over a number of years by financial support from the

National Research Council of Canada and discussions with many colleagues. I am grateful in particular to S.K. Banerjee, M.E. Evans, M. Fuller, R.T. Merrill and F.D. Stacey for debating, refuting and refining my ideas.

REFERENCES

AMAR, H., Magnetization mechanism and domain structure of multi-domain particles, *Phys. Rev.*, **111**, 149–153, 1958.

BANERJEE, S.K., On the origin of stable remanence in pseudo-single domain grains, *J. Geomag. Geoelectr.*, **29**, 319–329, 1977.

BEAN, C.P. and J.D. LIVINGSTON, Superparamagnetism, *J. Appl. Phys.*, **30**, 120S–129S, 1959.

BROWN, W.F., Virtues and weaknesses of the domain concept, *Rev. Mod. Phys.*, **17**, 15–19, 1945.

BROWN, W.F., Rigorous approach to the theory of ferromagnetic microstructure, *J. Appl. Phys.*, **29**, 470–471, 1958.

BROWN, W.F., *Micromagnetics*, pp. 143, Interscience, New York, 1963.

BUTLER, R.F. and S.K. BANERJEE, Theoretical single-domain grain size range in magnetite and titanomagnetite, *J. Geophys. Res.*, **80**, 4049–4058, 1975.

DAVIS, P.M. and M.E. EVANS, Interacting single-domain properties of magnetite intergrowths, *J. Geophys. Res.*, **81**, 989–994, 1976.

DAY, R., TRM and its variation with grain size: a review, *J. Geomag. Geoelectr.*, **29**, 233–265, 1977.

DE BLOIS, R.W. and C.D. GRAHAM, Domain observations on iron whiskers, *J. Appl. Phys.*, **29**, 931–939, 1958.

DICKSON, G.O., C.W.F. EVERITT, L.G. PARRY, and F.D. STACEY, Origin of thermoremanent magnetization, *Earth Planet. Sci. Lett.*, **1**, 222–224, 1966.

DUNLOP, D.J., Magnetite: behavior near the single-domain threshold, *Science*, **176**, 41–43, 1972.

DUNLOP, D.J., Thermoremanent magnetization in submicroscopic magnetite, *J. Geophys. Res.*, **78**, 7602–7613, 1973a.

DUNLOP, D.J., Theory of the magnetic viscosity of lunar and terrestrial rocks, *Rev. Geophys. Space Phys.*, **11**, 855–901, 1973b.

DUNLOP, D.J., Superparamagnetic and single-domain threshold sizes in magnetite, *J. Geophys. Res.*, **78**, 1780–1793, 1973c.

DUNLOP, D.J., The hunting of the 'psark' (abstract), *EOS (Trans. Am. Geophys. Union)*, **57**, 904, 1976a.

DUNLOP, D.J., Thermal fluctuation analysis: a new technique in rock magnetism, *J. Geophys. Res.*, **81**, 3511–3517, 1976b.

DUNLOP, D.J., Magnetic hysteresis of single-domain and two-domain iron oxide particles, in preparation, 1977.

DUNLOP, D.J. and M-M. BINA, The coercive force spectrum of magnetite at high temperatures: evidence for thermal activation below the blocking temperature, *Geophys. J. R. Astron. Soc.*, 1977 (in press).

DUNLOP, D.J. and E.D. WADDINGTON, The field dependence of thermoremanent magnetization in igneous rocks, *Earth Planet. Sci. Lett.*, **25**, 11–25, 1975.

DUNLOP, D.J. and G.F. WEST, An experimental evaluation of single-domain theories, *Rev. Geophys. Space Phys.*, **7**, 709–757, 1969.

DUNLOP, D.J., F.D. STACEY, and D.E.W. GILLINGHAM, The origin of thermoremanent magnetization: contribution of pseudo-single-domain magnetic moments, *Earth Planet. Sci. Lett.*, **21**, 288–294, 1974.

EVANS, M.E., Single domain oxide particles as a source of thermoremanent magnetization, *J. Geomag. Geoelectr.*, **29**, 267–275, 1977.

FLETCHER, E.J. and W. O'REILLY, Contribution of Fe^{2+} ions to the magnetocrystalline anisotropy constant K_1 of $Fe_{3-x}Ti_xO_4$ ($0 < x < 0.1$), *Proc. Phys. Soc. (London), Solid State Phys.*, **7**, 171–178, 1974.

FREI, E.H., S. SHTRIKMAN, and D. TREVES, Critical size and nucleation field of ideal ferromagnetic particles, *Phys. Rev.*, **106**, 446–455, 1957.

GALT, J.K., Motion of a ferromagnetic domain wall in Fe$_3$O$_4$, *Phys. Rev.*, **85**, 664–669, 1952.

KNELLER, E.F. and F.E. LUBORSKY, Particle size dependence and remanence of single-domain particles, *J. Appl. Phys.*, **34**, 656–658, 1963.

McNAB, T.K., R.A. FOX, and A.J.F. BOYLE, Some magnetic properties of magnetite (Fe$_3$O$_4$) microcrystals, *J. Appl. Phys.*, **39**, 5703–5711, 1968.

MERRILL, R.T., The demagnetization field of multidomain grains, *J. Geomag. Geoelectr.*, **29**, 285–292, 1977.

NÉEL, L., Théorie du traînage magnétique des ferromagnétiques en grains fins avec applications aux terres cuites, *Ann. Géophys.*, **5**, 99–136, 1949.

NÉEL, L., Théorie du traînage magnétique des substances massives dans le domaine de Rayleigh, *J. Phys. Radium*, **11**, 49–61, 1950.

NÉEL, L., Some theoretical aspects of rock magnetism, *Adv. Phys.*, **4**, 191–242, 1955.

OZIMA, M. and M. OZIMA, Origin of thermoremanent magnetization, *J. Geophys. Res.*, **70**, 1363–1369, 1965.

PARRY, L.G., Magnetic properties of dispersed magnetite powders, *Philos. Mag.*, **11**, 303–312, 1965.

SCHMIDT, V.A., A multidomain model of thermoremanence, *Earth Planet. Sci. Lett.*, **20**, 440–446, 1973.

SHIVE, P.N., Dislocation control of magnetization, *J. Geomag. Geoelectr.*, **21**, 519–529, 1969.

SHTRIKMAN, S. and D. TREVES, Internal structure of Bloch walls, *J. Appl. Phys.*, **31**, 147S–148S, 1960.

SOFFEL, H.C., Domain structure of titanomagnetites and its variation with temperature, *J. Geomag. Geoelectr.*, **29**, 277–284, 1977.

STACEY, F.D., A generalized theory of thermoremanence, covering the transition from single domain to multidomain magnetic grains, *Philos. Mag.*, **7**, 1887–1900, 1962.

STACEY, F.D., The physical theory of rock magnetism, *Adv. Phys.*, **12**, 45–133, 1963.

STACEY, F.D. and S.K. BANERJEE, *The Physical Principles of Rock Magnetism*, pp. 195, Elsevier, New York, 1974.

STEPHENSON, A., The observed moment of a magnetized inclusion of high Curie point within a titanomagnetite particle of lower Curie point, *Geophys. J. R. Astron. Soc.*, **40**, 29–36, 1975.

STONER, E.C. and E.P. WOHLFARTH, A mechanism of magnetic hysteresis in heterogeneous alloys, *Philos. Trans. R. Soc. (London), Ser. A*, **240**, 599–642, 1948.

STRANGWAY, D.W., E.E. LARSON, and M. GOLDSTEIN, A possible cause of high magnetic stability in volcanic rocks, *J. Geophys. Res.*, **73**, 3787–3795, 1968.

STREET, R. and J.C. WOOLLEY, A study of magnetic viscosity, *Proc. Phys. Soc. (London), Ser. A*, **62**, 562–572, 1949.

VERHOOGEN, J., The origin of thermoremanent magnetization, *J. Geophys. Res.*, **64**, 2441–2449, 1959.

WOHLFARTH, E.P., The remanent magnetization of haematite powders, *Philos. Mag.*, **46**, 1155–1164, 1955.

Adv. Earth Planet. Sci., **1**, 87–97, 1977

On the Origin of Stable Remanence in Pseudo-Single Domain Grains*

S.K. BANERJEE

*Department of Geology and Geophysics, University of Minnesota,
Minneapolis, U.S.A.*

(Received June 23, 1977)

The existence of pseudo-single domain (PSD) grains was postulated by F.D. Stacey in 1962 but a viable quantitative model for the magnetic moment of the canonical case of a PSD grain (a two domain grain with a single 180° wall) was explicitly derived only in 1974 by F.D. Stacey and S.K. Banerjee. According to this model, the net magnetic moment of a PSD grain arises from the wall moment alone and the high stability to alternating field demagnetization is mainly due to pinning of the wall by large surface anisotropy in such fine grains. Recent theoretical work in micromagnetics shows not only that the canonical case described above is a crude approximation but that the complexities of spin orientations in such grains are far too large so as to allow a theoretical prediction of the expected moments. Much more hope, however, can be gained from recent experimental studies on rare earth-cobalt alloys and yttrium iron garnet crystals. These studies support the basic contention that the role of surface anistropy is indeed the predominant one in explaining the high coercivity of PSD grains. These studies also provide useful experimental models for future research in this area in rock magnetism.

1. Introduction

The term 'pseudo-single domain grain' (PSD grain) was coined by STACEY (1962) to describe a group of hypothetical magnetite grains which were larger than the critical size threshold (d_0) for single domain behavior and yet were difficult to demagnetize using high peak alternating fields (AF), e.g., 1,000–2,000 Oe. Assuming $d_0 = 300$ Å to be true for magnetite, Stacey postulated that grains between 300 Å and 50,000 Å ($= 5\,\mu$m) will carry net spontaneous magnetizations (although, in general, decreasing with increasing grain size) due to discreteness of the domain wall Barkhausen jumps in such fine grains and he attributed this physical mechanism in PSD grains to be responsible for the 'non-demagnetizable remanences' or 'random moments' observed by IRVING *et al.* (1961) in some rocks. However, in a note added in proof to his 1962 paper, Stacey mentioned that B.J. Patton had found no evidence of a 'non-demagnetizable remanence' when he carried out AF demagnetizations in a carefully controlled dc field-free space. In a later paper STACEY (1963) first repeated Patton's personal communication to him, then pointed out that IRVING *et al.*'s (1961)

---

* Presented as an invited contribution to the 1976 National Fall Meeting of the A.G.U.

observed 'non-demagnetizable remanence' was not aligned with the sample elongation as expected for PSD moments arising from Barkhausen discreteness effect. He therefore concluded that '...from the theoretical viewpoint, the possibility must be allowed that random moments represent an intrinsic property of at least some rocks, although the balance of evidence is against this conclusion...'. Three years later in 1966, PSD moments due to Barkhausen discreteness were revived by Dickson *et al.* (1966) who used then recently obtained thermoremanent magnetization (TRM) data on sized magnetite grains by Parry (1965) and fitted the TRM data for grains between 1 μm and 17 μm in size (d) to a theoretical formula for TRM of PSD grains which led to an approximate d^{-1} dependence. This is quite unlike the d-independent behavior observed for the larger grains and easily explained by a multi-domain theory of TRM.

In the same paper, Dickson *et al.* (1966) further stated that the observed d^{-1} dependence of TRM for PSD grains could not be accounted for by Verhoogen's (1959) magnetoelastic model of domain wall-pinning by internal defects nor by Ozima and Ozima's (1965) modification of Verhoogen's model in which they proposed that the defect-ridden regions are separated from other grains by grain boundaries. The chief reason for discarding the magnetoelastic model and domain-wall pinning was that even if a magnetite grain has internal non-uniform stress (σ) equal to its breaking stress (σ_{max}), the calculated bulk magnetoelastic energy ($E_\sigma = \frac{3}{2}\lambda_s\sigma_{\mathrm{max}}$ where λ_s = average magnetostriction constant) will still be one order of magnitude smaller than the magnetostatic energy expected from a single domain (SD) grain with an approximate demagnetizing factor of 4.

By the early 1970's the reality of PSD grains and their expected properties had been discovered. The high AF-stabilities of their TRMs were similar to that of true SD grains but their saturation remanence ratios (J_r/J_s) for dc fields in excess of 10,000 Oe were in the range of 0.1 to 0.3, distinctly lower than the 0.5 value expected from the Stoner-Wohlfarth theory for true SD grains. Dunlop (1973) replotted the then available TRM versus d data (his Fig. 7) on a log-log plot and confirmed Dickson *et al.*'s (1966) contention that from about 0.2 μm to about 50 μm grain size low field TRM is proportional to d^{-1}. However, TRM for grain size below 0.1 μm was less than that expected from a d^{-1} dependence and even showed a decrease with decreasing grain size. Additionally, Dunlop (1973) pointed out that although rigorous theoretical calculations or experimental observations of domain wall thickness (t) for magnetite grains below $d=0.1$ μm were not available, it is likely that the wall itself will occupy a very large fraction of the grain volume and the Barkhausen discreteness (a typical 'bulk' property) model may not be applicable to these extremely small PSD grains. Instead, Dunlop suggested that a 'wall-like' or 'wave-like' domain structure was present in these PSD grains leading to a non-uniform spin structure and hence $J_r/J_s < 0.5$. Hand in hand with this went the high AF-stability of the TRM since '...demagnetizing fields and not local barriers to wall motion will limit any magnetization changes...' (Dunlop, 1973).

Meanwhile, Stacey and Banerjee had come to a similar conclusion by April,

1972, when the completed manuscript of their book (STACEY and BANERJEE, 1974) was submitted to the publisher. Realizing that the d^{-1} dependence was strongly suggestive of a surface effect (surface/volume ratio of a sphere$=d^{-1}$) and incorporating some of the ideas of VERHOOGEN (1959) and OZIMA and OZIMA (1965), STACEY and BANERJEE (1974) offered an alternative PSD model, dependent strongly on domain wall pinning by surface defects. They further added precision to DUNLOP's (1973) suggestion of a 'wave-like' structure by calculating the maximum wall magnetization ($m_{max}=7.6\times10^{-14}$ gauss) expected from a PSD grain which was almost completely filled with a 180° domain wall. DUNLOP et al. (1974) attempted an experimental test of the Stacey-Banerjee model. Their observations showed that the model predicted a value of m_{max} which was smaller than that observed. A second and important observation was that sub-micron magnetite grains did *not* have the same (normalized) TRM dependence with applied *dc* field, a feature which could be predicted from two approximations made by STACEY and BANERJEE (1974) viz., (1) that there will be a distribution of surface defects on each PSD grain and the moment contribution from these regions would vary from zero to m_{max} and (2) that these distributions would not vary with grain size in the PSD region, i.e., *d*-values ranging from sub-micron to, say, 20 μm (above which 'bulk' or true multi-domain behaviour predominates).

The above background is necessary to appreciate what follows: (a) the importance of recent theoretical and experimental inroads made in rock magnetism and pure magnetism in order to understand better the PSD phenomenon in two-domain grains, and (b) the pertinent recent advances made in fine particle magnetism based on a new and better understanding of surface physics.

2.   Canonical Case of PSD—180° Domain Wall Moment

Since the goal of this paper is, in part, to emphasize the relevant advances in pure magnetism rather than rock magnetism, there will be little rock magnetism *per se* in this section. Figure 1 is taken from STACEY and BANERJEE (1974) which incorporates much excellent data obtained through courtesy of L.G. Parry. Figure 1 shows clearly that around $d\leq20$ μm, TRM slowly increases with decreasing *d*-value (following a d^{-1} or PSD behavior) and although the exact cut-off point for separating PSD behavior from bulk behavior may not be exactly 20 μm, the previously mentioned log-log plot of similar data by DUNLOP (1973) leaves one in no doubt that TRM acquisition in low field does change at around $d=20$ μm. DAY (1977, this issue) has claimed a $d^{-0.7}$ dependence but the experimental data are not yet good enough to distinguish 0.7 from 1.0. Dunlop's figure also shows that below $d\sim0.1$ μm the experimentally observed TRM values are consistently lower than the Stacey-Banerjee model for which $m_{max}=7.6\times10^{-14}$ gauss. Merely changing the value of m_{max} downwards to fit the data is not satisfactory since we are still left with the other discrepancy with the model, that the normalized TRM induction curves do not overlap for different grain sizes in the PSD range. The only conclusion that can be drawn is that the

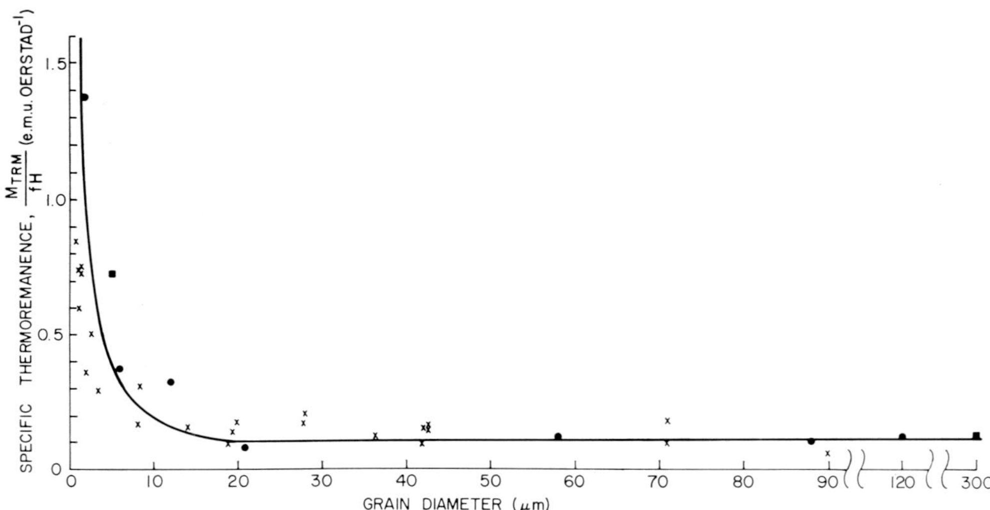

Fig. 1.   TRM in 1 Oe of a sample containing 1 vol. % magnetite versus grain size (after Stacey and Banerjee, 1974).

Stacey-Banerjee model of 1974 is only a first-order approximation and that we have to accept that different sized grains in the PSD range have different magnitudes of PSD moments. It is my belief that an experimental approach is likely to be much more successful than a theoretical one to construct a more realistic model.

Although it is *reasonable* to assume that a 180° domain wall *will* occupy a large volume of a small two-domain (2D) grain, Fig. 2 (Butler and Banerjee 1975) provides theoretical evidence from a rigorously calculated model that it is indeed so for pure magnetite. For magnetite cubes whose edges range in size from $d = 0.072\ \mu$m (720 Å) to $d = 0.082\ \mu$m (820 Å), the domain wall thickness t varies only from 425 Å to 455 Å. The theoretical d_0 was deduced by Butler and Banerjee (1975) to be 765 Å; thus the domain wall in such a grain would indeed occupy 60% or more of the grain volume, making it a canonical example of a true PSD grain or a 'psark' (Dunlop, 1977, this issue). Butler and Banerjee used this new value of domain wall thickness to recalculate $m_{\max}$ of Stacey and Banerjee (1974) and found comfort in the fact that the new $m_{\max}$ was closer to the experimental value of $m_{\max}$ observed by Dunlop *et al.* (1974). The agreement, however, could be purely fortuitous because of the comments about the other discrepancy mentioned earlier in the preceding paragraph.

As if to emphasize the above point, some recent work in the area of micro-magnetics makes it almost futile to try to refine the $m_{\max}$ value as long as we hold on to the first-order Stacey-Banerjee model of a 2D grain, which is also used by Dunlop (1977, this issue). LaBonte (1969) and Hubert (1969) have carried out rigorous two-dimensional micromagnetic calculations to predict the complete magnetization state in a 2D thin magnetic film. While LaBonte used an approach based on Monte

Carlo trials, Hubert has been successful in solving the problem analytically. They both find that the slowly rotating spin structure inside a 180° domain wall is asymmetrically placed with reference to the domain wall boundaries, thus making it very

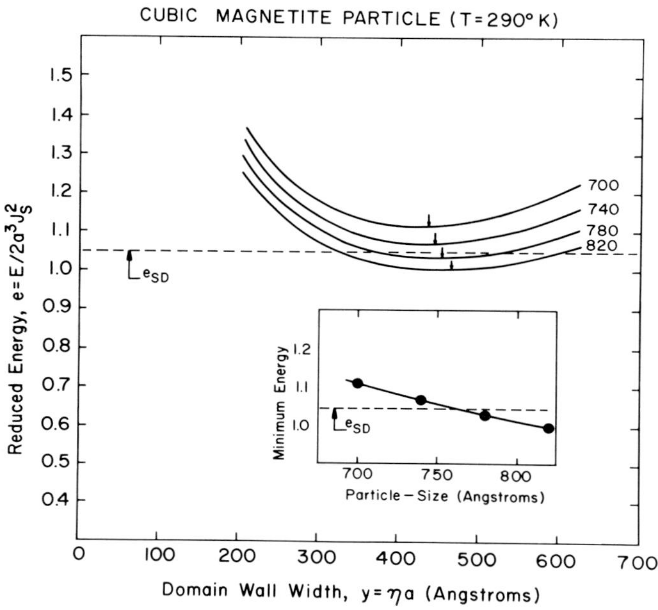

Fig. 2.   Reduced energy versus domain wall width for magnetite cubes of four indicated grain sizes (in Ångstroms). The inset shows how d_0 was determined (after BUTLER and BANERJEE, 1975).

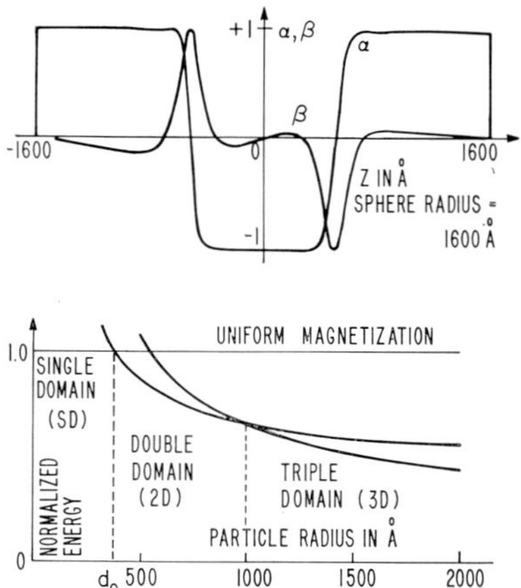

Fig. 3.   Upper part=variations in local magnetization directions (α in the plane of the paper and β at right angles) in a three domain sphere of cobalt. Lower part=normalized energy versus grain size for the same sphere indicating thresholds for two and three domain states (after STAPPER, 1969).

difficult to calculate analytically the value of $m_{\max}$ in a 2D grain. These results obtained for thin films might also be applicable to fine particles if allowance is made for the lateral dimension because the thicknesses of the films are comparable to the grain size under our consideration. Since an appeal to analogy may not be the best argument, let us look at the upper part of Fig. 3 which describes STAPPER's (1969) theoretical attempts to provide a micromagnetic model for, not a thin film, but a spherical three domain grain of cobalt. The same paper also describes the 2D case and in the lower part of Fig. 3, the grain size thresholds for 2D and three domain (3D) cases are displayed. The variations of the direction cosines (α, β) of the magnetic moment in the upper part show that on either side of the 180° domain wall, the main domains do not have the expected uniform orientations, resulting in what STAPPER (1969) calls 'skirts' or departures which, of course, will complicate theoretical calculations for the $m_{\max}$ expected from a pure 2D grain. A more interesting result is shown on the bottom of Fig. 3. Rock magnetists often wonder how large a grain size region is controlled by the canonical 2D arrangement, before the 3D situation is obtained, which will then result in a decrease of the net wall moment of the grain, if not a complete cancellation due to two oppositely directed wall moments. If the case with cobalt is any guide, the answer is only about 500 Å. This makes it very difficult to imagine that true 2D grains (or psarks à la DUNLOP, 1977) will ever be present in any appreciable number, in turn making it unlikely that an experimental search for psarks will be rewarded very soon.

3.   PSD Properties Due to Surface Anistropy

Let us therefore look at the latest researches in the field of fine particle magnetism and the role played by surface anisotropy in such grains. As with many other areas in magnetism, NÉEL (1953) provided an early insight into the problem of magnetic surface anistropy experienced by the top layer of atoms in an infinite

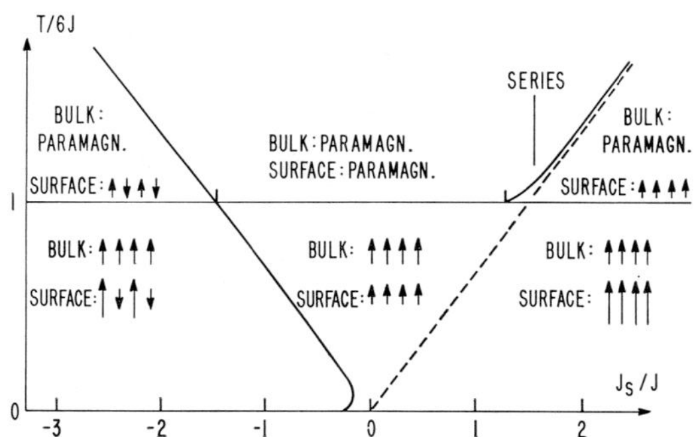

Fig. 4.   Magnetic phase diagram in the T-J_s plane where T=absolute temperature and J_s=surface exchange integral. J=bulk ferromagnetic exchange integral. The magnetic orders are indicated with arrows (after BINDER and HOHENBERG, 1976).

half-space. The classical dipolar interaction was used as the basis of the model and the presence of free space at the surface (i.e., the missing layer of magnetic atoms above) causes a non-cancellation of the microscopic dipolar anistropy terms when integrated over a layer which is only a few atoms thick. BINDER and HOHENBERG (1976) have carried out Monte Carlo (again!) computations for such two-dimensional systems based on Ising and Heisenberg models. Their first numerical results are presented as a phase diagram in Fig. 4 where the surface layer is allowed to have an exchange constant different from that of the bulk material, both in magnitude and sign. By varying these parameters, Binder and Hohenberg have been able to show how sensitively the temperature variation of bulk magnetization depends on the local conditions at the surface. One of the more dramatic conclusions of their analysis is shown on the left hand side of the diagram (normalized surface exchange constant=large and negative). Here an antiferromagnetic surface magnetization is present above the Curie point of the ferromagnetic bulk material. On cooling, a strongly ferrimagnetic layer develops at the surface which must affect the magnetization (and TRM) of the bulk material.

A result which will perhaps be more easily appreciated by rock magnetists has only recently been reported by VLASKO-VLASOV et al. (1976). In Fig. 5 I have reproduced their observations of the domain structure in Yttrium Iron Garnet (YIG). YIG is cubic in crystal structure like magnetite and it also has [111] as its easy axis. In Fig. 5, therefore, the reader is looking at 180° domain-walls in the (110) plane and, interestingly enough, the oppositely magnetized wall moments are indeed separated by Bloch lines (see DUNLOP, 1977, this issue, for a detailed description of a Bloch line). Vlasko-Vlasov et al. call them Néel lines but a clear decision cannot be made from observations in only one plane. For me, the most interesting part of Fig. 5 lies in the broadening of the walls at the surface due to surface anisotropy which explains how relatively thin walls in the bulk region can be strongly modified and pinned at the surface. The surface expressions of these walls were studied by these authors using magnetic colloids.

Questions ought to be raised as to the relative importance of wall pinning by

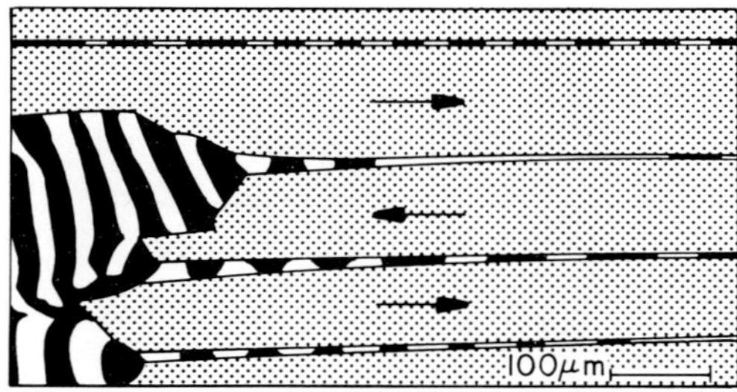

Fig. 5.   Magnetic domain structure and domain wall fine structure in YIG as revealed by the magneto optic effect. The stippled regions are the main domains, the black and white structure inside the 180° domain walls represent Bloch lines (after VLASKO-VLASOV et al., 1976).

internal defects inside bulk material versus pinning at the surface due to Néel-type surface defects such as dislocation pile-ups. The truth is that so far as magnetite is concerned, despite DICKSON *et al.*'s (1966) theoretical arguments, we still do not know the real magnitude of surface pinning energy of domain walls in magnetite. For one thing, Dickson *et al.*, used the spontaneous magnetization of *single domain* (SD) grains to show that the magnetostatic energy is likely to be one order of magnitude greater than the maximum bulk magnetoelastic energy. Since the calculation of magnetostatic energy incorporates the square of the spontaneous magnetization, their calculation is inappropriate since PSD grains possess only a fraction of the spontaneous magnetization of SD grains. Secondly, in the light of what we now know about the relatively large surface anisotropy of small grains, it is questionable whether average bulk magnetostriction constants should have been used to calculate the magnetoelastic energy of the grain.

However, some recent experiments in the field of rare earth-cobalt magnets may provide a model which could be successfully used in rock magnetism in order to discern the origin of high intrinsic coercivity (and AF stability of TRM) in fine grained magnetite. The most studied compounds have been the samarium-cobalt alloys, $SmCo_5$ and $SmCo_7$. The microscopic coercivities expected from these compounds on the basis of intrinsic magnetocrystalline anisotropy are in excess of 100,000 Oe. However, the best synthesized material, with or without doping with minor amounts of other elements, have yielded coercivities that are less than 60,000 Oe. The explanations which have been proffered are based on two competing mechanisms for magnetization reversal: (a) nucleation model and (b) wall motion model. In the nucleation model, it is claimed that an apparently saturated grain still retains a few small reversely magnetized or randomly oriented surface domains whose existence is

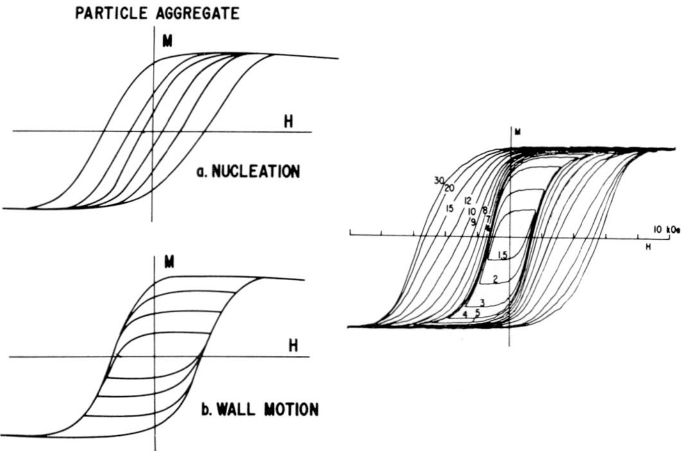

Fig. 6. Left hand side: both upper and lower loops indicate hysteresis data obtained with increasing H_{max} values. The upper loops indicate control of coercivity by surface anisotropy, the lower ones indicate control by bulk defects. Right hand side: actual data from a copper modified $SmCo_7$ sample with increasing H_{max} (after BECKER, 1976).

controlled by predominant local surface anisotropy due to the presence of corners, holes or other defects. However, since the total contribution to the magnetic moment of such small regions is negligible, hysteresis loops measured in increasingly large maximum fields ($H_{\max}$) should show the idealized behavior depicted in the top left-hand section of Fig. 6. Here the loops get fatter along the H or field axis since the application of increasing $H_{\max}$ succeeds in aligning more and more of the high co-ercivity surface domains. The model is termed 'nucleationmodel' because these same surface domains serve as the first nucleation sites for new reverse domains when magnetization reversal is attempted after a true and complete saturation in the highest possible field. The behavior expected from the wall motion model is exemplified in the bottom left hand side of the same figure. If the high coercivity regions are internal and not located at the surface then the overall coercive force is determined by the statistical distribution of the internal defects (dislocation pile-ups, foreign atomic cluster precipitates, etc.) and is invariant with $H_{\max}$. However, as greater and greater applied fields sweep out the domain walls through the bulk of the grain, the value of saturation magnetization increases, resulting in the predicted loops. On the right hand side of Fig. 6, data obtained with a copper-modified $SmCo_7$ material by BECKER (1976) are presented. It shows clearly that even in the same material, as $H_{\max}$ is increased the dominant barrier to a change in magnetization changes from bulk pinning of the wall motion to pinning at the high surface anisotropy regions (nucleation centers).

We can draw two conclusions from the right hand side of Fig. 6: (a) in $SmCo_7$ at least, surface anisotropy-constrolled domains have higher stabilities than the defect controlled ones in the bulk sample and (b) that although it is difficult to understand the exact orientations of the magnetic spins at the boundary of a surface domain and into the bulk of the grain proper, it appears that the net magnetization arising from the bulk of the grain can be changed or reversed *without* modifying or annihilating the small surface domains. This was the general picture suggested by STACEY and BANERJEE (1974, e.g., Fig. 10.4) but Becker's actual observations put it on a firm experimental ground.

4. A Forward View

If one incorporates at least some of the ideas and observations from the area of fine particle magnetism into rock magnetism, certain conclusions and future avenues for research become apparent.

1) The canonical case of the PSD model for magnetite, i.e., a 2D grain with one $180°$ wall filling most of the grain volume may indeed be a psark and elude detection by experiments. A rigorous calculation of its expected individual moment $m_{\max}$ may be beyond our abilities, given the recent complexities unravelled by micromagnetics.

2) Even though a 2D grain may be hard to synthesize or theorize about, it is abundantly clear that local fluctuations in surface anisotropy can provide a very

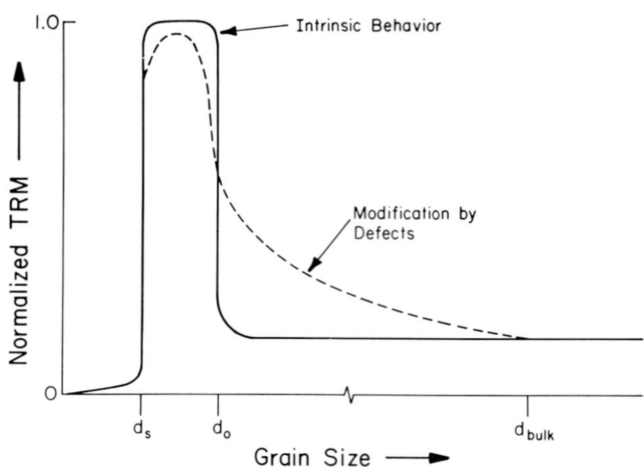

Fig. 7. A cartoon describing the author's present understanding of the influence of defects on the intrinsic TRM versus grain size behavior. The PSD region extends from d_0 to d_{bulk}.

satisfactory alternative (and something that we already know to be true from Fig. 1)—that is, there is a range of grain sizes between d_0 and d_{bulk} (i.e., the threshold above which bulk behavior or true multi-domain behavior predominates) which have PSD-type properties, low J_r/J_s ratios compared to SD grains coexisting with high AF stability such as is seen in SD grains. In this connection, it may be relevant to draw the reader's attention to a cartoon, which is Fig. 7. The basic model owes its origin to a suggestion by Kronmüller and Hilzinger (1973) regarding the coercivity of rare earth cobalt alloys (incidentally, they, in turn, refer to A. Aharoni as being the source of the original idea). The solid line refers to the TRM as ideally expected from a grain size increase from d_s (superparamagnetic threshold) through d_0 (single to two domain threshold) to d_{bulk} and beyond. However, as soon as defects are introduced into the grain, either sizable bulk defects or surface defects (I prefer the latter), the dashed curve results. Above d_0, TRM and AF stability both *rise* due to PSD effect (cf., Merrill (1975), Fig. 4) whereas between d_0 and d_s, TRM *falls* because magnetization reversal need not be a coherent rotation process, leading to a slight decrease in TRM and a more visible *decrease* in AF stability. Only future experiments will show the validity or otherwise of this cartoon.

My understanding of the PSD effect has been aided by the help and critical advice obtained over the years from D.J. Dunlop, R.T. Merrill, F.D. Stacey and J. Verhoogen. More recently, in Minnesota, I have been fortunate enough to work out new ideas with my colleagues, R.F. Butler, S. Levi and Emeritus Professor W.F. Brown, Jr. Professor Brown is also responsible for drawing my attention to the papers on micromagnetics and for kindly providing me with a copy of his own English translation of the recent paper by Vlasko-Vlasov *et al.* I also thank the National Science Foundation and the National Aeronautics and Space Administration for their financial support of my work (NSF grant nos. DES 75-21796 and OCE 76-15255 and NASA grant no. NGR 24-005-248).

REFERENCES

BECKER, J.J., Reversal mechanism in copper-modified cobalt-rare-earths, *IEEE Trans. Magn.* **12**, 965–967, 1976.

BINDER, K. and P.C. HOHENBERG, Magnetic ordering and critical behavior near surfaces, *IEEE Trans. Magn.* **12**, 66–79, 1976.

BUTLER, R.F. and S.K. BANERJEE, Theoretical single-domain grain size range in magnetite and titanomagnetite, *J. Geophys. Res.*, **80**, 4049–4058, 1975.

DAY, R., TRM and its variations with grain size, *J. Geomag. Geoelectr.*, **29**, 233–265, 1977.

DICKSON, G.O., C.W.F. EVERITT, L.G. PARRY, and F.D. STACEY, Origin of thermoremanent magnetization, *Earth Planet. Sci. Lett.*, **1**, 222–224, 1966.

DUNLOP, D.J., Superparamagnetic and single-domain threshold sizes in magnetite, *J. Geophys. Res.*, **78**, 1780–1793, 1973.

DUNLOP, D.J., The hunting of the 'Psark', *J. Geomag. Geoelectr.*, **29**, 293–318, 1977.

DUNLOP, D.J., F.D. STACEY, and D.E.W. GILLINGHAM, The origin of thermoremanent magnetization: Contribution of pseudo-single-domain magnetic moments, *Earth Planet. Sci. Lett.*, **21**, 288–294, 1974.

HUBERT, A., Stray-field-free magnetization configurations, *Phys. Status Solidi*, **32**, 519–534, 1969.

IRVING, E., P.M. STOTT, and M.A. WARD, Demagnetization of igneous rocks by alternating magnetic fields, *Philos. Mag.*, **6**, 225–241, 1961.

KRONMÜLLER, H. and H.R. HILZINGER, The coercive field of hard magnetic materials, *Int. J. Magn.*, **5**, 27–30, 1973.

LABONTE, A.E., Two-dimensional Bloch-type domain walls in ferromagnetic films, *J. Appl. Phys.*, **40**, 2450–2458, 1969.

MERRILL, R.T., Magnetic effects associated with chemical changes in igneous rocks, *Geophys. Surv.*, **2**, 277–311, 1975.

NÉEL, L., L'anisotropie superficielle des substances ferromagnétiques, *C.R.*, **237**, 1468–1470, 1953.

OZIMA, M. and M. OZIMA, Origin of thermoremanent magnetization, *J. Geophys. Res.*, **70**, 1363–1369, 1965.

PARRY, L.G., Magnetic properties of dispersed magnetite powders, *Philos. Mag.*, **11**, 303–312, 1965.

STACEY, F.D., A generalized theory of thermoremanence covering transition from single domain to multidomain magnetite grains, *Philos. Mag.*, **7**, 1887–1900, 1962.

STACEY, F.D., The physical theory of rock magnetism, *Adv. Phys.*, **12**, 45–133, 1963.

STACEY, F.D. and D.K. BANERJEE, *Physical Principles of Rock Magnetism*, Elsevier, Amsterdam, 1974.

STAPPER, C.H., Micromagnetic solution for ferromagnetic spheres, *J. Appl. Phys.*, **40**, 798–802, 1969.

VERHOOGEN, J., The origin of thermoremanent magnetizations, *J. Geophys. Res.*, **65**, 2441–2449, 1959.

VLASKO-VLASOV, V.K., L.M. DEDUKH, and V.I. NIKITENKO, Domain structure of monocrystals of Yttrium Iron Granet, *Zh. Eksp. Teor. Fiz.*, **71**, 2291–2304, 1976 (in Russian).

Adv. Earth Planet. Sci., **1**, 99–112, 1977

The Preparation, Characterization and Magnetic Properties of Synthetic Analogues of Some Carriers of the Palaeomagnetic Record

J.B. O'DONOVAN and W. O'REILLY

Department of Geophysics and Planetary Physics, School of Physics,
The University of Newcastle upon Tyne, Newcastle upon Tyne,
United Kingdom

(Received June 2, 1977)

Synthetic analogues of two magnetic mineral systems have been prepared: (i) a titanomagnetite system $Fe_{2.4-\delta}Al_\delta Ti_{0.6}O_4$ ($0 < \delta < 0.45$). (ii) a titanomaghemite system prepared by oxidation of $Fe_{2.4-\delta}Mg_\delta Ti_{0.6}O_4$ ($0 < \delta < 0.35$). The following properties were determined: unit cell edge, Curie temperature, thermomagnetic behaviour, saturation magnetization, isothermal remanence and coercive force. The introduction of Al, Mg and vacancies into titanomagnetites reduces saturation magnetization and coercive force. Curie temperature is lowered by the presence of Al and Mg ions but rises with increasing concentration of vacancies.

1.   Introduction

The mechanisms by which rocks and minerals may acquire and preserve a record of the palaeogeomagnetic field have been the subject of many reviews. Essential to each mechanism is the presence, at the time of formation of the rock, of a perturbation which, when combined with a weak ambient field, results in a remanence (TRM, CRM, DRM) resistant to later changes in the field unless the perturbation is reapplied. The development of mathematical models (e.g., NÉEL, 1955; MERRILL, 1976) to describe these mechanisms forms a major part of the physical foundation of the palaeomagnetic method. Of equal importance is the identification of the remanence-carrying minerals in rocks and comparison of the observed characteristics of such minerals with those required by the theoretical models (e.g., DUNLOP *et al.*, 1973).

In the case of igneous rocks the remanence carriers are titanomagnetites ($Fe_{3-x}Ti_xO_4$, $0 < x < 1$) or compounds derived by 'substitution', oxidation and/or unmixing processes such as sub-solvus exsolution or inversion (e.g., HARGRAVES and PETERSEN, 1971). The true titanomagnetite is presumably never found in nature but it is usual to apply the term to near-stoichiometric Fe-Ti spinel oxides with up to, perhaps, 10% of other cation species (of which Mg and Al are the most abundant together with Mn, Cr and Zn). The degree of non-stoichiometry is limited at high temperature ($\gtrsim 400°C$) (e.g., HAUPTMAN, 1974) and the products of such ('deuteric') oxidation beyond the monophasic limit, are intergrowths containing a spinel component. At low temperature and under favourable circumstances, e.g., those apparently holding in the submarine crust (HALL, 1976), the degree of non-stoichiometry is not limited. The monophasic spinel products of low temperature oxidation are referred to as titanomaghemites and the oxidation process itself as maghemitization.

An understanding of the magnetic effects of maghemitization is required for the interpretation of the measured NRM intensities of samples collected from the ocean floor (e.g., RYALL and ADE-HALL, 1975) and certain aspects of the anomalies in the geomagnetic field observed over the ocean basins. A time dependence of intrinsic magnetic properties (and hence intensity and stability of remanence) may result not only from maghemitization but also from ionic reordering at ambient temperatures. This latter effect may be especially enhanced by the presence of Mg ions. Indeed, from a survey of ferrite literature, it seems probable that the intrinsic properties of titanomaghemites will be significantly modified by the presence of the minor cation species. Changes in chemical composition may also significantly shift the critical volumes separating the superparamagnetic/stable monodomain/multidomain regimes (e.g., BUTLER, 1973). Finally, changes in microstructure due, for example, to inversion of a metastable titanomaghemite will lead to a modification of the intensity and stability of remanence.

The object of the present study (which forms part of a continuing systematic investigation of the titanomagnetites and their derivatives) is to determine some of the basic magnetic properties of two systems. The first is a titanomagnetite subsystem based on $Fe_{2.4}Ti_{0.6}O_4$ in which the minor cation species are represented by Al ions in concentrations up to 0.45 cations per formula unit. The second is a titanomaghemite system, again based on $Fe_{2.4}Ti_{0.6}O_4$ containing Mg ions as representatives of the minor cation species. The properties to be measured include Curie temperatures, thermomagnetic curves and the hysteresis parameters, saturation magnetization, isothermal remanence and coercive force.

Over the past two decades (from, e.g., AKIMOTO *et al.* (1957)) the preparation of synthetic analogues corresponding more and more closely to naturally occurring materials, and the laboratory simulation of processes believed to take place in nature, have made a significant contribution to progress in (i) the identification of remanence carrying minerals, (ii) determining their genesis and characteristics, and (iii) establishing the physical mechanisms by which the stable remanence was acquired and retained. The present investigation is in the tradition of such studies.

2. Sample Synthesis and Analysis

The starting materials were analytical grade Al_2O_3, MgO, TiO_2, Fe_2O_3 and Fe metal. The Al_2O_3, TiO_2 and MgO were dehydrated by heating at 600°C, the Fe was freshly reduced in pure hydrogen at 380°C and the Fe_2O_3 was heated to 650°C in a stream of dry oxygen to convert any Fe^{2+} to Fe^{3+}, to dehydrate the oxide and to enhance the solid state reactivity of the material. To ensure some measure of constant grain size and to eliminate the formation of conglomerates, the starting materials were sieved using fine nylon meshes before mixing in stoichiometric proportions.

The homogeneity of the mixture was ensured by violent shaking in a mechanical agitator for about an hour. This was preferred to ball-milling under acetone because the iron, being malleable, when ground in a ball-mill tends to form flakes which opposes homogeneity. The mixture was pressed into pellets under pressures of approximately 2,700 kgm/cm² and placed in alumina boats in a mullite tube. The tube was evacuated and flushed with argon, the process being repeated four times over a period of several hours and the tube finally filled with argon to a pressure of 200 torr so that at the firing temperature, 1,350°C, the internal pressure would be slightly in excess of atmospheric. A stream of pure nitrogen was passed over the outside of the tube during the six-hour firing period and subsequent cooling.

The pellets were then crushed, repressed and the firing procedure carried out a second time, the argon now being replaced by a mixture of 90 volume percent high

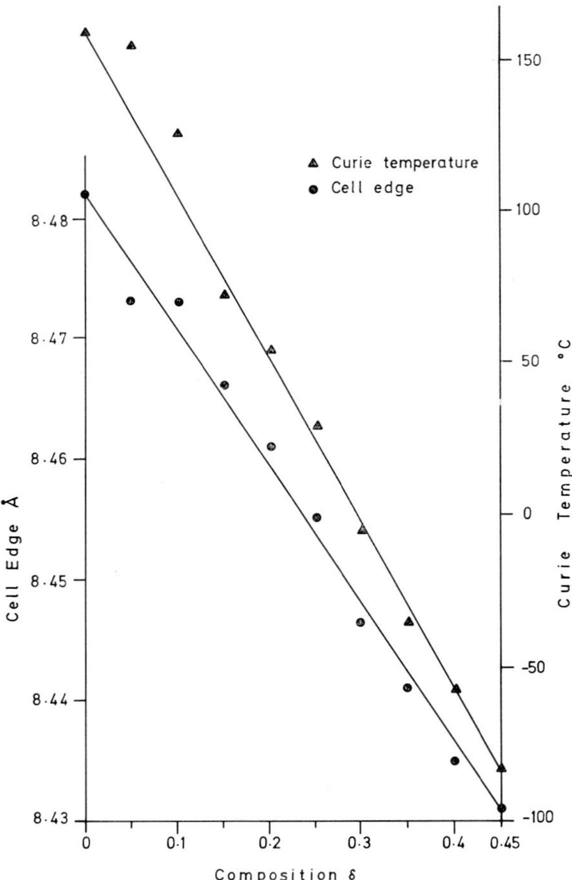

Fig. 1. Variation of Curie temperature and cell edge with composition in the system $Fe_{2.4-\delta}Al_\delta Ti_{0.6}O_4$. The Curie temperatures were obtained using a low field susceptibility bridge and the cell edges determined from Debye-Scherrer powder photographs.

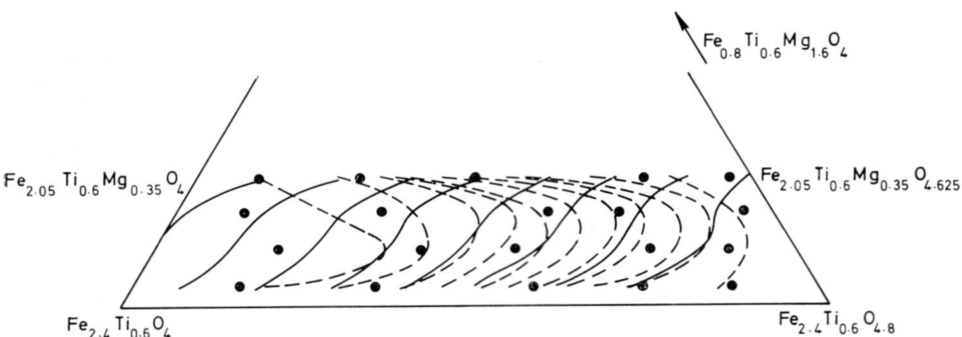

Fig. 2.   The preparation and characterization of the titanomaghemite system based on Fe$_{2.4}$-Ti$_{0.6}$O$_4$ with up to 0.35 Mg^{2+} ions per formula unit. The chemical compositions of the twenty synthesized samples are represented by the solid circles.   Contours of unit cell edge of the fine grain materials, determined from Guinier-de Wolff powder photographs are shown as dashed lines.   The left hand contour corresponds to 8.47 Å and that on the right hand to 8.37 Å. Intermediate contours are at 0.01 Å intervals.   Curie point contours, shown as solid lines, were determined using samples sealed in evacuated capsules.   From left to right the contours correspond to 160°C rising in intervals of 40°C to 440°C.

purity argon and 10 volume percent CO_2-CO mixture in a ratio slightly in excess to that appropriate to Fe$_{2.4}$Ti$_{0.6}$O$_4$ at 1,350°C.   The excess made the mixture slightly reducing.   The CO-CO_2 mixture both buffers the oxygen partial pressure and significantly promotes the rate of transfer of oxygen between the grains.   X-ray analysis, chemical analysis and Curie point determination showed that single phase near-stoichiometric material had been produced.

Samples of the magnesium substituted series were prepared for maghemitization by ball-milling for 12 hours in a slurry of high purity methanol.   Oxidation was carried out in air at temperatures between 200 and 350°C in a thermobalance.   The compositions of the twenty titanomaghemite samples so produced were determined by a combination of wet chemical analysis and thermogravimetry (O'Donovan and O'Reilly, 1977a).

Cell edge and Curie temperature data for the synthesized systems are shown in Figs. 1 and 2.

3.   Magnetic Measurements

3.1   Room temperature magnetic hysteresis loops

Hysteresis loops which yielded saturation magnetization, saturation isothermal remanence and coercive force were obtained using a vibrating sample magnetometer of the Foner type with a maximum field of 13.5 kOe.   The titanomagnetite and titanomaghemite powders were dispersed in a calcium fluoride matrix (oxide : matrix = 1 : 5) and pressed into pellets.   A summary of the hysteresis data is given in Tables 1a and 1b and some typical chart recordings of hysteresis loops in Fig. 3.

Table 1a.   Summary of hysteresis loop parameters for the system $Fe_{2.4-\delta}Al_\delta Ti_{0.6}O_4$.

Composition δ	Coercive force H_c (Oe)	Saturation magnetization M_s (emu/g)	Remanence M_{rs} (emu/g)	M_{rs}/M_s
0.05	40	27.3	3.28	0.15
0.1	30	23.5	2.35	0.10
0.15	23	17.8	1.60	0.09
0.2	22	14.5	1.58	0.19
0.25	14	10.4	0.97	0.09

The data was obtained using powders sieved to obtain a grain size range 37–44 μm and dispersed in a non-magnetic matrix.

Table 1b.   Summary of hysteresis loop parameters for the system $Fe_{(2.4-\delta)R}Mg_{\delta R}Ti_{0.6R}\square_{3(1-R)}O_4$ ($0<\delta<0.35$).

Composition parameters		Coercive force H_c (Oe)	Saturation magnetization M_s (emu/g)	Remanence M_{rs} (emu/g)	M_{rs}/M_s
δ	z				
0.05	0.16	2,000 (1,540)	22.6 (29.3)	12.9 (15.6)	0.57 (0.53)
	0.36	2,020	20.3	11.6	0.57
	0.59	1,890	19.9	9.93	0.50
	0.75	720	13.8	7.61	0.55
	0.89	290 (355)	11.7 (15.5)	6.55 (5.88)	0.56 (0.38)
0.15	0.20	1,760 (1,210)	19.1 (26.4)	9.74 (12.7)	0.51 (0.48)
	0.42	1,700	16.8	9.21	0.55
	0.57	1,560	15.8	8.70	0.55
	0.78	645	13.8	6.62	0.49
	0.90	260 (385)	12.7 (26.1)	6.22 (10.2)	0.49 (0.39)
0.25	0.12	1,350	17.6	8.96	0.51
	0.35	1,400	18.5	9.64	0.52
	0.62	930	14.7	7.64	0.52
	0.74	620	14.4	7.22	0.50
	0.95	165 (355)			0.48 (0.42)
0.35	0.12	1,060	17.6	8.63	0.49
	0.30	1,170	16.3	8.17	0.50
	0.51	955	13.2	7.27	0.55
	0.81	270	14.2	6.68	0.47
	0.97	155 (345)	13.2 (29.0)	6.33 (11.9)	0.48 (0.41)

The parameter R describes the degree of non-stoichiometry of the spinel compounds, where $R=8/(8+z(1.6-\delta))$ and z is the fraction of the Fe^{2+} in the stoichiometric compound oxidized to Fe^{3+}. The data was obtained from dispersed ball-milled powders with particle size of the order of 1,000 Å.

Data in brackets refer to samples which have been annealed in vacuo (low z values) or samples which have been inverted (high z values).

3.2   Temperature dependence of saturation magnetization

The variation of saturation magnetization between 80 and 900K was determined using a magnetic balance. The main features of the curves are first the P-type behaviour (i.e., a magnetization-temperature curve passing through a maximum) and, in the more non-stoichiometric samples, complex irreversible behaviour above about 400°C. Examples of the former are shown in Fig. 4 and of the latter in Fig. 5.

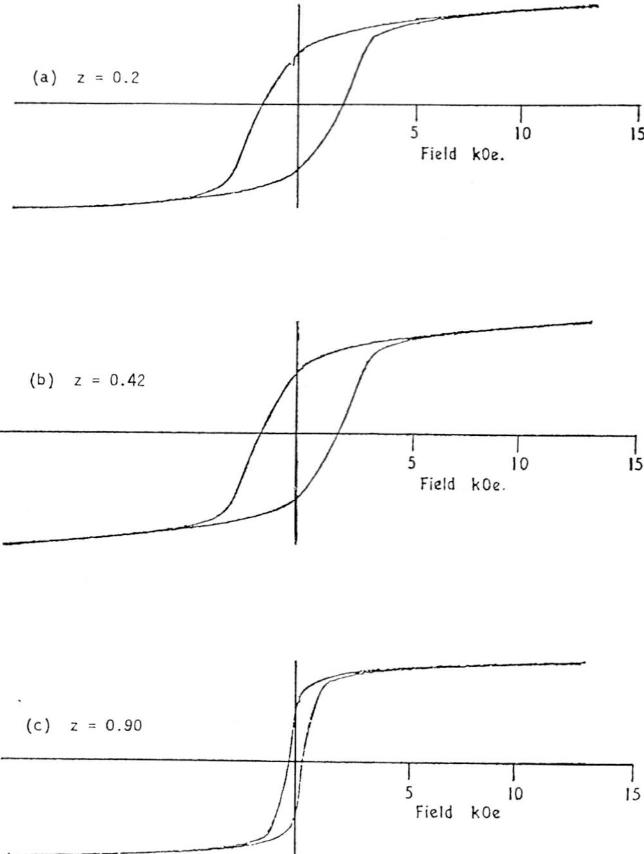

Fig. 3. Representative chart recordings of room temperature hysteresis loops. These were obtained for samples produced by oxidation of finely ground $Fe_{2.25}Mg_{0.15}Ti_{0.6}O_4$ to compositions given by $z=0.2$, 0.42 and 0.90 (z, the oxidation parameter, is the fraction of initial Fe^{2+} converted to Fe^{3+}).

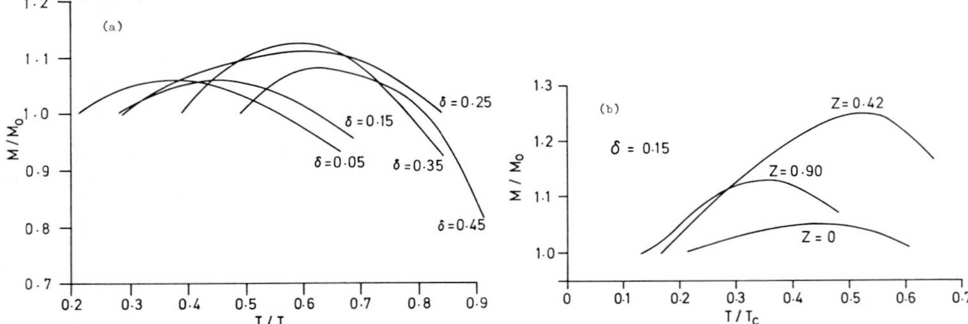

Fig. 4. (a) Thermomagnetic curves of members of the series $Fe_{2.4-\delta}Al_\delta Ti_{0.6}O_4$ (values of δ indicated on the curves) in the temperature range 80–300K plotted in reduced units. The P type behaviour becomes more pronounced as Al concentration increases. (b) Thermomagnetic curves of the titanomaghemite series derived from $Fe_{0.25}Mg_{0.15}Ti_{0.6}O_4$ (values of z indicated on the curves) in the temperature range 80–300K plotted in reduced units. The P type behaviour becomes at first more, and then less, pronounced as maghemitization proceeds.

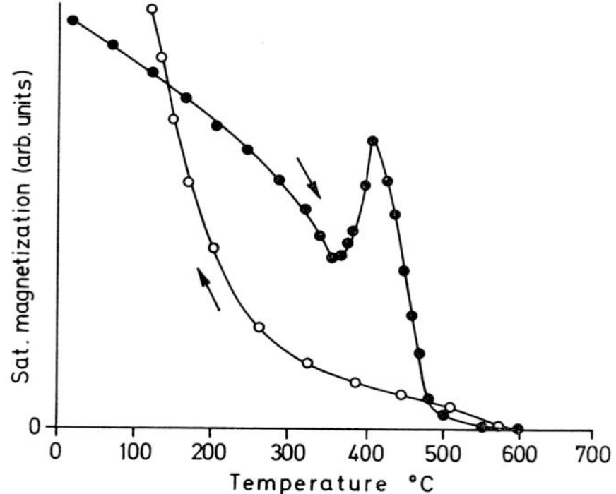

Fig. 5. Thermomagnetic curve for magnesium substituted titanomaghemite, encapsulated in vacuo, in the temperature range 20–600°C. The rise in magnetization at about 350°C indicates that the temperature limit of the stability field for low temperature non-stoichiometry has been reached and the sample is inverting to a multiphase intergrowth without change in bulk chemical composition. The final assemblage, after cooling from 600°C, seems to be reached by a two-stage process.

4.   Discussion

4.1   Cell edge measurements

The interionic distances for particular cation-anion pairs in a given coordination are approximately constant in a wide range of fluorides and oxides (e.g., SHANNON and PREWITT, 1969).   An alternative expression of this observation is found in the assignment of definite values to ionic radii and the 'rigid sphere' model of solids.   It is therefore possible, in principle, to calculate a cell edge for a spinel compound if the cation distribution is known.   However a survey of the literature (e.g., BLASSE, 1964) shows there to be many spinel series with markedly non-linear saturation magnetization-composition curves but with approximately linear cell edge-composition curves.   This is also true for the titanomagnetite solid solution for which cell edges calculated using cation distributions inferred from saturation magnetization data are not in agreement with those actually measured (PRÉVOT and POIX, 1971). The concept of the invariance of cation-anion distance may therefore be of only limited usefulness in the case of spinels.   It might be expected for example, that the degree of covalency (leading to a 'squeezing' of the ions) of a particular cation in octahedral sites may be a function of the radii of the nearest neighbour tetrahedral site population.

The variation of cell edge with composition in the present synthesized systems is, in a general way, consistent with accepted ionic radii.   Thus the replacement of $0.45\,Fe^{3+}$ ions by Al^{3+} brings about a marked reduction in cell edge ($\sim 0.6\%$), whereas the substitution of an equal amount of Mg^{2+} for Fe^{2+} results in a fall of only $\sim 0.1\%$. Maghemitization reduces the cell edge due to the conversion of Fe^{2+} to Fe^{3+}.   The size of vacancies must be considered as uncertain due to the absence of any significant quantity of cell edge and u-parameter data for non-stoichiometric spinels. The Prévot-Poix interionic distances, which were conceived within the context of the titano-magnetites, seem in better agreement with the present data than the Shannon-Prewitt distances, obtained from a large number of compounds of different structures.

The cell edges of the Al-substituted titanomagnetite system were determined using the standard Nelson-Riley extrapolation for the back reflections of Debye-Scherrer powder photographs. The uncertainty in the determination is of the order 0.002 Å. The cell edges of the Mg-substituted titanomaghemite system were deter-mined from Guinier-de Wolff powder photographs using silicon as an internal stand-ard. This technique was adopted for the titanomaghemite system first because it is particularly important to demonstrate the monophasic nature of the maghemitized oxidation products and secondly such fine grain material gives no back reflections in the Debye-Scherrer method. The uncertainty in the determinations is of the order of 0.006 Å. Cell edges determined by Guinier-de Wolff appear to be about 0.005 Å higher than those yielded by the Debye-Scherrer method for identical samples (Ö. Özdemir, private communication).

4.2 Curie temperatures

The Curie temperatures fall due to the introduction of diamagnetic ions and, as expected, the replacement of Fe^{3+} ions produces a greater fall than the replacement of an equal amount of Fe^{2+} ions (0.45 Al^{3+} ions per formula unit reduces the Curie temperature from about 160°C to -80°C, whereas 0.45 Mg^{2+} ions per formula unit reduces the Curie temperature to 110°C). Maghemitization increases the Curie temperature due to the conversion of Fe^{2+} to Fe^{3+} and the most highly oxidized compositions appear to have 'virtual' Curie temperatures in excess of 400°C.

Like unit cell edge, Curie temperatures of spinels depend on both composition and cation distribution and experimental values can be compared to those obtained from model calculations. The Molecular Field model suggests that for a given com-position the dependence on cation distribution is not strong, being proportional to the square root of the product of concentrations of the paramagnetic ions on the two sublattices (e.g., the calculated Curie temperatures for $Fe^{3+}(Mg^{2+}Fe^{3+})O_4$ and $Fe^{3+}_{2/3}Mg^{2+}_{1/3}(Fe^{3+}_{4/3}Mg^{2+}_{2/3})O_4$ would be in the ratio 1 : 0.95). Stephenson (1972) has derived the result for spinels containing two species of paramagnetic ion. Under the assump-tion of some plausible cation distribution such model calculations are in general agreement with the present data. Two comments which might be made about the calculations is that Molecular Field Theory overestimates the Curie temperatures in magnetically dilute materials when Molecular Field coefficients or exchange constants have been evaluated in concentrated materials. Secondly the cell edge-based 'distance correction' previously applied in such calculations (Stephenson, 1972; Readman and O'Reilly, 1972) would seem incompatible with the concept of invariance of metal-oxygen distances. Solid solutions in which exchange constants, evaluated from as-ymptotic Curie temperatures, are found to vary with composition as in, for example, the ferromagnesian olivines (HOYE and O'REILLY, 1972) may indeed indicate the limitations of the concept of metal-oxygen distance invariance.

4.3 Thermomagnetic curves

A number of mechanisms have been proposed to account for the P-type behaviour of titanomagnetites. Some models involve the exchange interactions (SCHULT, 1970; STEPHENSON, 1972) others invoke special properties of Fe^{2+} ions in octahedral symmetry (BANERJEE *et al.* 1967; ONO and CHANDLER, 1968). In the systems studied in the present investigation (i) the P-type behaviour becomes more pronounced as Al concentration increases, (ii) the P-type behaviour becomes at first more, and then less, pronounced as maghemitization proceeds. The second result would seem to exclude the latter models (special properties of Fe^{2+}) at least as applicable to titanomaghemites.

Of the former models the present data for the titanomaghemites may be consistent with the Néel mechanism. Briefly, this requires a positive (ferromagnetic) BB intrasublattice interaction. The stronger this interaction, the more pronounced the P-type behaviour will be. At low z the Fe^{2+} content of octahedral (B) sites is likely to be less than the Fe^{3+} content. At high z the Fe^{3+} content of B sites will be greater than the Fe^{2+} content. At some intermediate z the concentrations of the two types of ions may be equal. At this point the Fe^{2+}-Fe^{3+} BB interaction would make its maximum contribution to the total BB interaction. Although the Fe^{3+}-Fe^{3+} and Fe^{2+}-Fe^{2+} BB interactions are expected to be negative (BLASSE, 1964) it may be that the direct Fe^{2+}-Fe^{3+} BB interaction is positive and thus may provide a possible explanation for the observed behaviour of titanomaghemites.

N-type behaviour, and self-reversal of remanence, has been found in a basalt in which the magnetic constituent appears to be a titanomaghemite ($x\sim0.6$, $z\sim0.5$) containing some aluminium and magnesium (SCHULT, 1976). Synthetic samples have yet to reveal such effects and it may be that they are confined to a very narrow compositional range.

The shoulder or rise in magnetization found on the heating curves of encapsulated titanomaghemites indicates inversion to a multiphase intergrowth as the boundary of the region of low temperature non-stoichiometry is reached. The inversion temperatures (~300–$350°C$), at atmospheric pressure, are typical of those found in previous studies of maghemite and the titanomaghemites (e.g., READMAN and O'REILLY, 1970). The presence of magnesium and titanium does not therefore appear to stabilize the cation deficient spinel structure. Inversion of titanomaghemites appears to take place by a two-stage process (READMAN and O'REILLY, 1970; O'DONOVAN and O'REILLY, 1977a).

4.4 Magnetic hysteresis properties

The coercive forces of the Al-substituted system are typical of coarse grain sintered titanomagnetites at room temperature, although the M_{rs}/M_s ratios (~0.1–0.2) are rather higher than those reported in a previous study on sintered samples of grain size $>150\,\mu$m prepared by a purely self-buffering technique (DAY *et al.*, 1976). The coercive force also shows a systematic fall with increasing Al concentration. The coercive force in such (presumably) multidomain materials depends strongly on those elements of microstructure which are variable and indeterminate, such as diameter and volume fraction of inclusions and internal stress. The role of grain boundaries acting as obstacles to domain wall translation may not be important in the present samples (grain size 37–44 μm) which are probably of sub-single crystallite size. The present data may be fitted to the model expressions of NÉEL (1946) by the choice of plausible values for internal stress variation ($\sim 10^9$ dynes/cm^2), the volume fraction of material subject to a large disturbing stress (≤ 0.1) and the volume fraction of inclusions. The fall in coercive force with increasing Al concentration presumably reflects the variation of magnetostriction or magnetocrystalline anisotropy constants with chemical composition and cation distribution. The high values of the ratio M_{rs}/M_s may indicate the presence of fine particles in the 37–44 μm sieved fraction.

The hysteresis loops of the fine grain titanomaghemites (Fig. 3) are consistent with the properties expected for monodomain grains with uniaxial or mixed uniaxial and cubic anisotropies i.e., M_{rs}/M_s ratios $\gtrsim 0.5$ and, secondly, very high values of coercive force. A lower limit to the grain size can be obtained from X-ray line broadening measurements which, because of strain-broadening yield an underestimate of the grain size. Because of the possibility of flocculation, especially in magnetic materials, direct observation with the electron microscope may produce an overestimate of the grain size. The grain size of the titanomaghemite samples was considered to be of the order of 1,000 Å from a combination of both techniques. The values of the coercive force can be explained in terms of Stoner-Wohlfarth coherent rotation in which the dominant anisotropy is due to stress, with internal stresses of the order of 10^9 dynes/cm^2 and the magnetostriction coefficients of SYONO (1965) and KLERK *et al.* (1976). It seems probable that the major contribution to the magnetostriction coefficients comes from Fe^{2+} and hence the fall in coercive force with maghemitization. As laboratory maghemitization requires prolonged heating at low temperatures ($\leq 300°$C) other alternative or contributory causes for the fall in coercive force are annealing out of the stresses induced by the ball-milling and re-ordering of Mg^{2+} ions. The effect of annealing some samples in vacuo was studied, but a more comprehensive and systematic investigation would be necessary to fully understand the effects of annealing. The results are shown in Table 1b. The sample $\delta = 0.05$, $z = 0.16$ was heated for two days at 250°C and $\delta = 0.15$, $z = 0.2$ for two and a half days at 425°C. The coercive forces decreased by 22% and 30% respectively while the saturation magnetization increased in both cases by about 30%. This would seem to indicate that the major decrease in coercive force with maghemitization is not due to annealing. A number of the high z samples were inverted by heating to 650°C in vacuo. In all cases the coercive force increased significantly (see Table 1b) and the M_{rs}/M_s ratios fell. This would be consistent with the production of an intergrowth containing iron-rich spinel with shape anisotropy. Interactions between the magnetic regions within the grain may have reduced both coercive force and M_{rs}/M_s ratios in this case as both of these parameters are lower than values obtained for similar intergrowths produced by laboratory simulation of deuteric oxidation (O'DONOVAN and O'REILLY, 1977b).

4.5 Some applications of studies of synthetic analogues

4.5.1 Identification of magnetic minerals

The measured characteristic properties of synthetic analogues are a valuable aid in identifying the magnetic minerals in rock samples and locating their positions on the FeO-Fe_2O_3-TiO_2 ternary diagram or more generally in 'composition space'. In naturally occurring spinel minerals with a multiplicity of cation species, the composition parameters (x, z) must be carefully defined. Data of the kind shown in Fig. 1 has been applied to the identification of minerals in subaerial basalts (CREER and IBBETSON, 1970), submarine basalts (RYALL and ADE-HALL, 1975) and submarine sediments (KENT and LOWRIE, 1974). The magnetic consequences of inversion are also qualitative indicators of maghemitization. However in previous investigations (OZIMA and SAKAMOTO, 1971; READMAN and O'REILLY, 1972) the effect of the presence of cation species other than Fe and Ti has not been included. Thus degrees of maghemitization inferred from a combination of electron probe microanalysis, for example, and a comparison of Curie temperature measurement and/or X-ray analysis with the catalogue of data available from synthetics must be subject to large uncertainties. The data obtained in the present study should contribute to the reduction of such uncertainties.

4.5.2 The study of mineralogical processes

A comparison of the characteristics of the magnetic fraction extracted from submarine pillow basalts of different ages from the Mid-Atlantic Ridge crest at $45°$N and the properties of synthetic analogues has enabled the evolution of the magnetic minerals in such basalts, to their observed state in the collected rock samples, to be traced out (RYALL and ADE-HALL, 1975; MARSHALL and COX, 1972). The pillows were thus found to show a radial variation in degree of maghemitization, the exterior being the most oxidized. The degree of maghemitization was also found to increase from the youngest to the oldest pillows. It appears therefore that the fine grain material produced in such quenched samples, which would be particularly suitable as a starting material for maghemitization, acquires oxygen from and/or loses cations to the sea water. The rate of maghemitization observed in the basalts is consistent with the reaction rate at room temperature (substantial oxidation in $\sim 10^6$ years) obtained from laboratory simulation of maghemitization at temperatures of 200–$300°$C (READMAN and O'REILLY, 1970). The high degree of maghemitization observed in the oldest specimens (z, the 'fraction of reaction', ~ 0.8) is consistent with the results of the present investigation, in which the compositional stability field for titanomaghemites extends up to $z = 1$.

4.5.3 Intensity of magnetization in ocean basalts

The decrease of intensity of magnetization of submarine basalts with distance from a spreading centre appears to be well established (e.g., IRVING, 1970). The decrease of spontaneous magnetization with increasing degree of maghemitization (table 1b) will make a contribution to this effect (with a reduction perhaps approaching 50%) although not enough to account for the observed decrease ($\sim$90%). A further contribution will come from the shift of the superparamagnetic/stable monodomain boundary to larger volumes as the degree of maghemitization increases (BUTLER, 1973). The diameter of spherical particles at the superparamagnetic boundary is proportional to $(M_sH_c)^{-1/3}$. From the present data (table 1b) and taking a relaxation time of 100 seconds to represent the superparamagnetic/stable monodomain boundary the calculated critical diameters are 260 Å ($\delta=0.05$, $z=0.16$) rising to 350 Å ($\delta=0.35$, $z=0.12$) before maghemitization. After maghemitization the critical diameters are 620 Å ($\delta=0.05$, $z=0.89$) to 660 Å ($\delta=0.35$, $z=0.97$). Depending on the actual grain size distributions in submarine lavas the two mechanisms described above, together with intrinsic time-decay of remanence, would seem to provide a plausible description of the loss of remanence with time.

5.   Conclusions

The retrieval and assessment of the information recorded by basaltic rocks requires a knowledge, inter alia, of the intrinsic properties of the titanomagnetites and their derivatives. The following information has resulted from the present study:

1) The Curie temperature in the system $Fe_{2.4-\delta}Al_\delta Ti_{0.6}O_4$ falls linearly from 160°C ($\delta=0$) to -90°C ($\delta=0.45$). The Curie temperatures in the system $Fe_{(2.4-\delta)}Mg_{\delta R}Ti_{0.6R}\square_{3(1-R)}O_4$ ($R=8/[8+z(1.6-\delta)]$ and z is the fraction of reaction) fall with the introduction of diamagnetic Mg^{2+} ions and rise with maghemitization. The lowest Curie temperatures in the system are $\sim$150°C and the highest $\sim$500°C.

2) P type behaviour is observed throughout both systems. An enhanced P type behaviour follows from the introduction of Al^{3+} ions and maghemitization produces at first an enhancement and then a suppression of the P type behaviour.

3) The inversion temperatures for the titanomaghemite system studied lie in the range 300–400°C.

4) The room temperature saturation magnetization falls from about 32 emu/g for the unsubstituted, unoxidized, titanomagnetite to about 10 emu/g for $Fe_{2.15}Al_{0.25}Ti_{0.6}O_4$ and to about 12–13 emu/g for the most highly oxidized fine grain Mg substituted samples.

5) The hysteretic properties of the titanomagnetite system (grain size 37–44 μm) are consistent with the multidomain state. Those of the fine grain titanomaghemites (grain size $\sim$1,000 Å) are typical of monodomain grains ($M_{rs}/M_s \gtrsim 0.5$) with H_c in the range 2,000 ($z\sim0.1$) to 200 Oe ($z\sim0.9$).

This work forms part of a NERC sponsored research programme 'Thermoremanence in titanomagnetites'. One of the authors (J.B. O'Donovan) has been in receipt of a NERC studentship. We thank Dr. Z. Hauptman for invaluable assistance in the preparation of materials.

REFERENCES

AKIMOTO, S., T. KATSURA, and M. YOSHIDA, Magnetic properties of $TiFe_2O_4Fe_3O_4$ system and their change with oxidation, *J. Geomag. Geoelectr.*, **9**, 165–178, 1957.

BANERJEE, S.K., W. O'REILLY, T.C. GIBB, and N.N. GREENWOOD, Behaviour of ferrous ions in iron-titanium spinels, *J. Phys. Chem. Solids*, **28**, 1328–1335, 1967.

BLASSE, G., Crystal chemistry and some magnetic properties of mixed metal oxides with spinel structure, Philips Res. Rep., Suppl., No. 3, 1964.

BUTLER, R.F., Stable single domain to superparamagnetic transition during low temperature oxidation of oceanic basalts, *J. Geophys. Res.*, **78**, 6868–6876, 1973.

CREER, K.M. and J.D. IBBETSON, Electron microprobe analyses and magnetic properties of non-stoichiometric titanomagnetites in basaltic rocks, *Geophys. J.R. Astron., Soc.*, **21**, 485–511, 1970.

DAY, R., M.D. FULLER, and V.A. SCHMIDT, Magnetic hysteresis properties of synthetic titanomagnetites, *J. Geophys. Res.*, **81**, 873–880, 1976.

DUNLOP, D.J., J.A. HANES, and K.L. BUCHAN, Indices of multidomain magnetic behaviour in basic igneous rocks: alternating field demagnetization, hysteresis, and oxide petrology, *J. Geophys. Res.*, **78**, 1387–1393, 1973.

HALL, J.M., Does TRM occur in oceanic layer 2 basalts? *EOS, Abstr., Trans. Am. Geophys. Union.*, **57**, 906, 1976.

HARGRAVES, R.B., and N. PETERSEN, Notes on the correlation between petrology and magnetic properties of basaltic rocks, *Z. Geophys.*, **37**, 367–382, 1971.

HAUPTMAN, Z., High temperature oxidation, range of non-stoichiometry and Curie point variation of cation deficient titanomagnetite $Fe_{2.4}Ti_{0.6}O_{4+\gamma}$, *Geophys. J.R. Astron. Soc.*, **38**, 29–47, 1974.

HOYE, G.S. and W. O'REILLY, A magnetic study of the ferromagnesian olivines $(Fe_xMg_{1-x})_2SiO_4$, $0 < x < 1$, *J. Phys. Chem. Solids*, **33**, 1827–1834, 1972.

IRVING, E., The mid-atlantic ridge at 45°N XIV, Oxidation and magnetic properties; review and discussion, *Can. J. Earth Sci.*, **7**, 1528–1538, 1970.

KENT, D.V. and W. LOWRIE, Origin of magnetic instability in sediment cores from the central North Pacific, *J. Geophys. Res.*, **79**, 2987–3000, 1974.

KLERK, J., V.A.M. BRABERS, and A.J.M. KUIPERS, Magnetostriction of the mixed series $Fe_{3-x}Ti_xO_4$, *Int. Conf. Ferrites, Paris*, 1976.

MARSHALL, M. and A. COX, Magnetic changes in pillow basalt due to sea floor weathering, *J. Geophys. Res.*, **77**, 6459–6469, 1972.

MERRILL, R.T., The origin of TRM in multidomain grains, *EOS, Abstr., Trans. Am. Geophys. Union*, **57**, 905, 1976.

NÉEL, L., Bases d'une novelle theorie generalle du champ coercitif, *Cah. Phys.*, **25**, 19–44, 1946.

NÉEL, L., Some theoretical aspects of rock magnetism, *Adv. Phys.*, **4**, 191–243, 1955.

O'DONOVAN, J.B. and W. O'REILLY, Range of non-stoichiometry and characteristic properties of the products of laboratory maghemitization, *Earth Planet. Sci. Lett.*, **34**, 291–299, 1977a.

O'DONOVAN, J.B. and W. O'REILLY, Monodomain behaviour in multiphase oxidized titanomagnetite, *Earth Planet. Sci. Lett.*, **34**, 396–402, 1977b.

ONO, K., L. CHANDLER, and A. ITO, Mössbauer study of the ulvöspinel, Fe_2TiO_4, *J. Phys. Soc. Japan*, **25**, 174–176, 1968.

OZIMA, M. and N. SAKAMOTO, Magnetic properties of synthesized titanomaghemite, *J. Geophys. Res.*, **76**, 7035–7046, 1971.

PRÉVOT, M. and P. POIX, Un calcul du parametre cristallin des titanomagnetites oxydes, *J. Geomag. Geoelectr.*, **23**, 255–256, 1971.

READMAN, P.W. and W. O'REILLY, The synthesis and inversion of non-stoichiometric titanomagnetites, *Phys. Earth Planet. Inter.*, **4**, 121–128, 1970.

READMAN, P.W. and W. O'REILLY, Magnetic properties of oxidized (cation deficient) titanomagnetite $(Fe, Ti, \square)_3O_4$, *J. Geomag. Geoelectr.*, **24**, 69–90, 1972.

RYALL, P.J.C. and J.M. ADE-HALL, Radial variation of magnetic properties in submarine pillow basalts, *Can. J. Earth Sci.*, **12**, 1959–1969, 1975.

Schult, A., Effect of pressure on the Curie temperature of titanomagnetites $(1-x)Fe_3O_4 - xTiFe_2O_4$, *Earth Planet. Sci. Lett.*, **10**, 81–86, 1970.

Schult, A., Self reversal above room temperature due to N type magnetization in basalt, *J. Geophys.*, **42**, 81–84, 1976.

Shannon, R.D. and C.T. Prewitt, Effective ionic radii in oxides and fluorides, *Acta Cryst.*, B, **25**, 925–945, 1969.

Stephenson, A., Spontaneous magnetization curves and Curie points of spinels containing two types of magnetic ions, *Philos. Mag.*, **25**, 1213–1232, 1972.

Syono, Y., Magnetocrystalline anisotropy and magnetostriction of Fe_3O_4-Fe_2TiO_4 series, *Jpn. J. Geophys.*, **4**, 71–143, 1965.

Adv. Earth Planet. Sci., **1**, 113–122, 1977

Reduction of Hematite to Magnetite under Natural and Laboratory Conditions

Peter N. SHIVE and Jimmy F. DIEHL[*]

*Department of Geology, University of Wyoming,
Laramie, Wyoming, U.S.A.*

(Received June 20, 1977)

Thermomagnetic analysis of natural and synthetic samples containing hematite have confirmed SCHWARZ's (1969) observations that 1) extensive reduction of hematite occurs when the experiment is perfomed in vacuum but not when it is performed in air 2) the reduction is arrested when the sample is heated in air before the vacuum run. In addition, we have found that the reduction is controlled by backstreaming vacuum pump oils, which buffer the atmosphere in the furnace region, but is not dependent upon reducing agents within the samples. Heating of the sample creates a surface layer which is not permeable to the reducing atmosphere and which thus prevents reduction during subsequent heatings. X-ray diffraction data from reduced synthetic hematite shows that the reduced phase is magnetite. Reducing agents within samples can contribute to reduction, and in some cases are so important that reduction may occur in air. Reduction produces undesirable effects during thermal demagnetization procedures and paleointensity determinations on samples containing hematite, but these effects can be eliminated or controlled if certain precautions are followed.

1. Introduction

According to experimental work by BUDDINGTON and LINDSLEY (1964), hematite will be stable with respect to magnetite at temperatures of 500°C unless the oxygen fugacity is below 10^{-18} atmospheres. The oxygen fugacity would need to be even lower in order for magnetite to be formed at lower temperatures. Low oxygen fugacities are obviously present during cooling of many igneous rocks and have in fact been directly measured (GROMMÉ *et al.*, 1969). But for most sediments, particularly those at and near the surface of the earth, the atmosphere is sufficiently oxidizing that magnetite should be oxidized to hematite or maghemite and thus, if magnetite exists at all, it does so in unstable equilibrium with the atmosphere.

It is possible, however, for conditions in nature and/or in the laboratory to favor the reduction of hematite during heating. Because paleomagnetists are more accustomed to worry about the effects of oxidation of magnetite and titanomagnetite, hematite reduction can produce unexpected and unwelcome complications in standard analyses and experiments. The purpose of this paper is to examine some of the

---

[*] Now at Plant and Earth Science Department, University of Wisconsin, River Falls, WI 54022.

observations of hematite reduction, to establish the conditions favoring reduction and to discuss the implications of this process for paleomagnetism and rock magnetism.

2.   Reduction of Red Sediments

In 1969 Schwarz found that hematite was reduced during standard thermo-magnetic analysis of red sediments if the analysis was performed in a partial vacuum, but that the hematite was stable when the analysis was carried out in air. He also found that reduction in vacuum did not occur if the sample had first been heated to 700°C in air. He believed that magnetite was the reduced phase on the basis of Curie temperature and increased intensity of magnetization during the cooling cycle. He suggested that the samples contained enough organic material that the atmosphere during heating was sufficiently reducing to create magnetite from hematite.

We ran powdered samples from red units of the Casper Formation (Pennsyl-vanian-Permian) of southeastern Wyoming and the Ingleside Formation (Permian) of northern Colorado on a recording magnetic balance similar to that described by Doell and Cox (1967) and Schwarz (1968) and essentially duplicated Schwarz's (1969) results. Samples weighed about 250 mg and were measured in fields of about 10 kOe. Vacuums of 10^{-4} torr were achieved by use of a mechanical roughing pump and a water-cooled diffusion pump. Heating and cooling rates were about 15°C per minute. Figure 1 shows thermomagnetic curves for three samples (H1153, H1173, 17) run in vacuum. Thermomagnetic curves for these same samples in air indicate the presence of hematite, but no magnetite, in addition to some paramagnetic material. Beginning at temperatures as low as 350°C the curves show the generation of a more magnetic phase and the recovery of many times the initial magnetization on cooling. Figure 2 shows thermomagnetic curves of H1153 and 17 in air (2a, 2c respectively) followed by a run in vacuum (2b, 2d respectively). Initial heating in

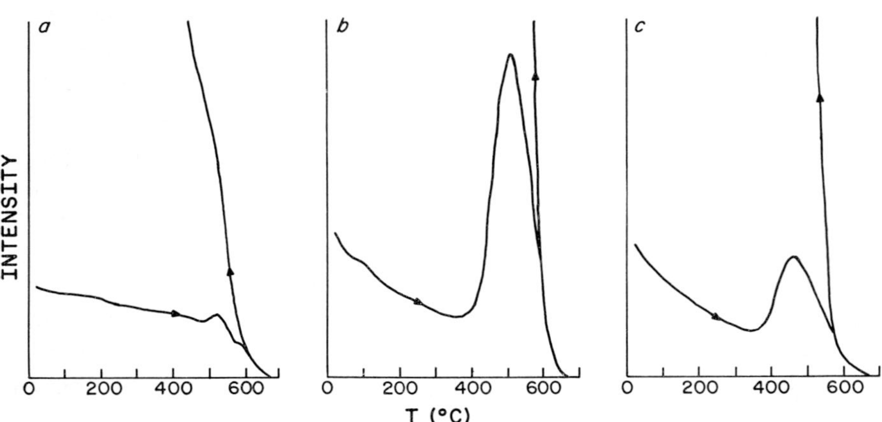

Fig. 1.   Thermomagnetic curves for samples (a) H1153, (b) H1173, and (c) 17 heated in vacuum without cold trap.

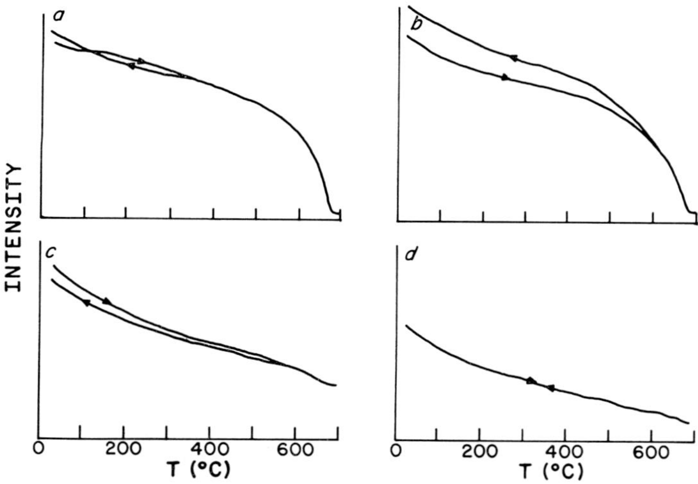

Fig. 2.   Thermomagnetic curves for samples H1153 and 17 (a, c, respectively) heated in air, and then subsequently heated in vacuum (b, d respectively) without cold trap.

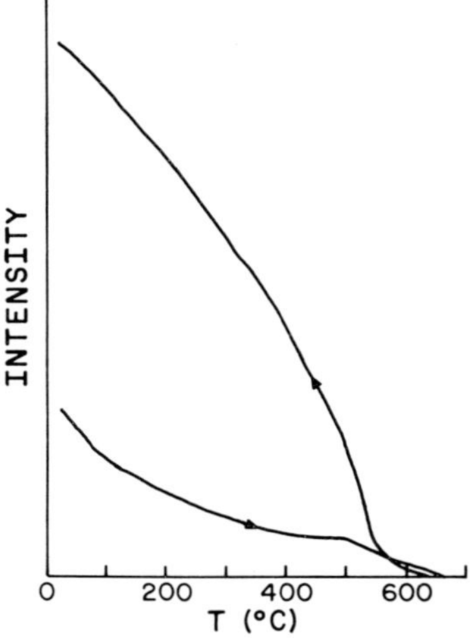

Fig. 3.   Thermomagnetic curve obtained from a specimen of sample 17 after heating in vacuum with cold trap. Vertical scale same as in Fig. 1c.

air suppressed reduction during the subsequent vacuum run, although a small amount of reduction did occur for sample H1153.

MARSHALL and COX (1972) found that the oxygen fugacity in their thermo-magnetic balance under diffusion pump vacuum was low enough that pure hematite was reduced at temperatures as low as 350°C. Apparently the vaccum pump oils backstream in to the furnace region and, since they are hydrocarbons, can produce a strongly reducing environment. In order to test the relative importance of the atmosphere of our vacuum system, we ran another specimen from sample 17 in vacuum with a liquid nitrogen cold trap (Fig. 3). The effect of the cold trap should be to reduce the amount of backstreaming of pump oils. Comparison of Fig. 3 with Fig. 1c shows that, although a significant amount of reduction occurred when a cold trap was used it was almost an order of magnitude less than when a cold trap was not used. It is also interesting to note the appearance of the different specimens of sample 17 which were subjected to different treatments. The specimen run in air retained its red color. The specimen run in vacuum with no cold trap was grey-black in color. The specimen heated in vacuum with a cold trap had a color between red and grey-black.

3.   Reduction of Artificial Hematite

Because the relative importance of the pump oils and possible reducing agents in the samples could not be separated in the experiments described above, we carried out a series of thermomagnetic analyses with artificial hematite. The artificial samples consisted of commercial hematite powder (99.999% Fe_2O_3) with a grain size of about 10 microns. When these samples were run in air the curves were reproducible. Figure 4 shows the synthetic hematite run in vacuum with (a) and without (b) cold trap. The curves have been normalized to have the same intensity after cooling. Comparison of the intensities before heating shows that the cold trap suppressed reduction by about a factor of two. Reduction was significant in both cases, but was in neither case as great as for the natural samples described above.

We also studied the influence of clay as a possible reducing agent. Differential thermal analysis of clays (JACKSON, 1956, p. 25; BRINDLEY and YOUELL, 1953; KERR *et al.*, 1949) shows that several stages of dehydration and the formation of oxides can take place below 700°C, and that these reactions occur at lower temperatures for the

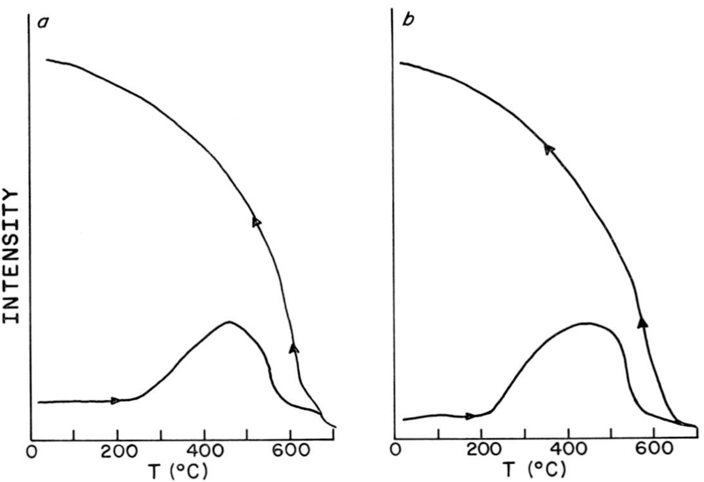

Fig. 4.   Thermomagnetic curves for synthetic hematite sample (a) heated in vacuum with cold trap and for a sample (b) heated in vacuum without could trap.

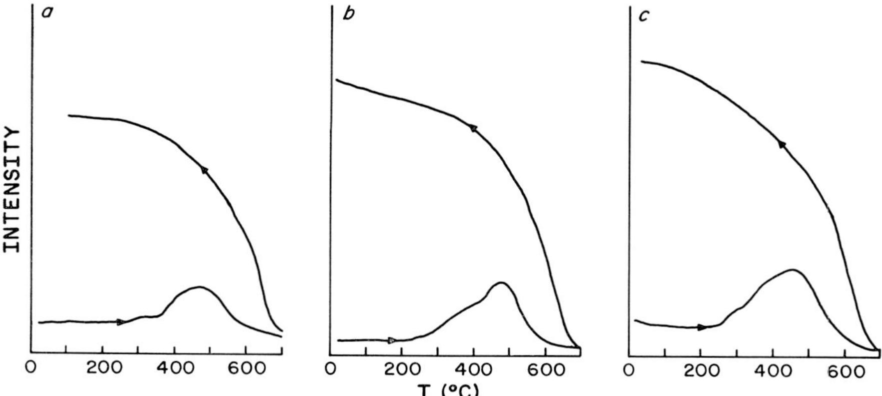

Fig. 5.   Thermomagnetic curves from samples of synthetic hematite mixed with (a) chlorite, (b) illite, and (c) nontronite.

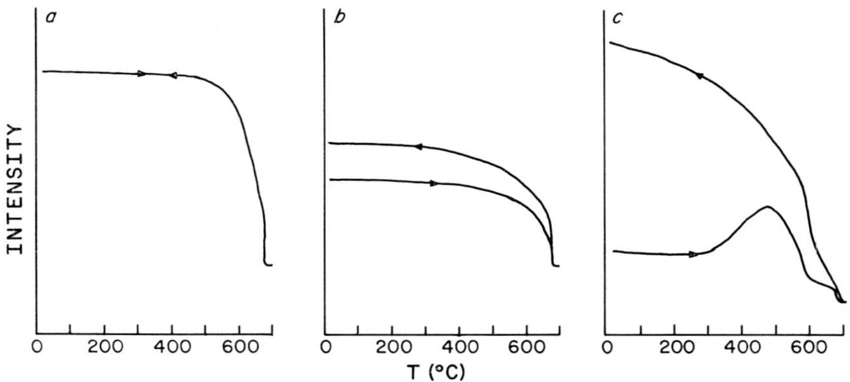

Fig. 6.   Thermomagnetic curves for a synthetic hematite sample (a) heated in air, (b) reheated in vacuum without cold trap, and (c) after being reground and reheated in vacuum without cold trap.

more iron rich clays. Figure 5 shows thermomagnetic curves for samples of artificial hematite powder mixed with equal amounts of the iron-bearing clays chlorite (5a), illite (5b), and nontronite (5c) in vacuum with cold trap. The curves are similar to each other and to the curve for synthetic hematite alone run under the same conditions (vacuum with cold trap). Thus the presence of iron-bearing clays does not affect the amount of reduction.

A final set of runs was made (Fig. 6) with synthetic hematite in air (6a), followed by a run with the same sample in vacuum without cold trap (6b). Then the sample was removed from the furnace, stirred up and rerun under the same conditions (6c). After heating in air and reheating in vacuum a very thin surface layer was blackened (the bulk of the sample was still red underneath this crust); apparently the small amount of reduction which did occur during this run was restricted to that surface. The nature of the surface is not known. It has a slight strength (for example, it supports the weight of the underlying powder when the bucket is turned over), which could result from partial fusion or from interlocking of newly formed magnetite crystals. After stirring, this surface layer is destroyed and reduction during the second vacuum run proceeded much as though the sample had not previously been heated in air.

4.   Composition of the Reduced Phase

Analysis of magnetic phases in these experiments is complicated because of the extremely small amounts of material involved. For example, our natural samples did not show any hematite in X-ray diffraction analysis although it was clearly present on the basis of magnetic behavior. Nor could we detect magnetite in the X-ray spectrum of natural samples after thermomagnetic runs in vacuum. Magnetite is suspected on the basis of Curie temperatures during cooling cycles and the strong increase in intensity.

Concentration of the reduced phase was achieved by repeatedly running synthetic hematite in vacuum and scraping off the reduced crust after each run. In addition to hematite peaks, the X-ray diffraction spectra show peaks corresponding to the (220), (400), and (333) lines of a spinel structure. The lattice parameter of the spinel is 8.39 Å $\pm$ 0.01 Å, which indicates that the phase is magnetite.

The Curie points observed during cooling of the synthetic samples (≈ 610°C) are higher than those observed for natural samples. This has led us to suggest (Shive and Diehl, 1977) that the magnetite lattice is under compression because (111) and (110) planes in magnetite try to conform to the more closely spaced (0001) and (1100) planes of the underlying hematite grain. Samara and Giardini (1969) showed that compression of magnetite causes its Curie temperature to increase at the rate of about 2°C per kilobar.

Curie points of about 600°C were also observed by Johnson et al. (1972) after heating synthetic hematite to 650°C in air in the presence of water vapor. X-ray diffraction data indicated the presence of a spinel with lattice parameter of 8.37 Å, and they suggested that the reduced phase was maghemite. Their hypothesis is attractive because water is known to stabilize the maghemite lattice (Elder, 1965) and because the lattice parameter and Curie temperature are about right for maghemite (Readman and O'Reilly, 1972) should it be able to exist at such high temperatures. We could not have produced maghemite since our reduction was carried out in vacuum and no water was present (although the natural samples may contain some water). In any case, our X-ray diffraction data indicates that magnetite is present in the reduced samples.

5.   Amount of Reduction

Comparative quantitative analysis of the thermomagnetic curves is difficult for several reasons. The amount of hematite in the natural samples is not known and is difficult to estimate from thermomagnetic curves because of contributions from paramagnetic phases. Also, the vacuums achieved during a run depended appreciably on the outgassing rate of the sample and on the condition of the vacuum chamber before the run. Finally, our vacuum gauge (Penning type ion gauge) does not measure oxygen fugacity and has somewhat different response to different gasses. Even with these problems, however, there are significant differences in the thermomagnetic curves and some useful information can be obtained by comparison.

The amount of hematite reduced can be estimated from the ratio of the sample magnetization before heating (due to hematite) to that after cooling (due to magnetite). For sample H1153, for which the initial amount of hematite can be most reliably estimated (Fig. 1a), about 5% of the hematite is reduced. This figure is probably much higher for samples 17 and H1173 since much of their initial intensity is due to paramagnetic material instead of hematite (Fig. 1b, 1c). For sample H1173 the amount of hematite reduced is probably well over 10%. The amount of reduction does not seem to depend on whether the samples are in the form of rock chips or powder magnetic separates.

These figures are substantially greater than the amount of synthetic hematite reduced under the same conditions (vacuum without cold trap), which is about 2%. This difference is probably due to grain size considerations. The hematite in the natural samples is mainly in the form of pigment and coatings around quartz and calcite grains. Thus it has a higher surface to volume ratio and can reduce more completely. Another difference between natural and artificial samples is that reduction of synthetic samples can begin below 200°C, much lower than for natural samples. We do not know why this should be the case, but it probably has little effect on the total amount of reduction since most of the reduction can occur above the Curie point of magnetite and thus not be apparent on the heating cycle.

The use of the cold trap inhibits reduction of both synthetic and natural samples. The amount of synthetic hematite reduced is down to about 1%. The effect of the cold trap on the reduction of sample 17 is even greater (compare Figs. 4 and 3).

6.   Discussion

Our experiments on synthetic hematite show that hematite will be reduced to magnetite at moderate temperatures under moderate (diffusion pump) vacuums, regardless of the presence of reducing agents in the sample. The only important factor controlling reduction of synthetic hematite, other than the presence or absence of a cold trap in the vacuum line, is whether or not the samples have been first heated in air. This process inhibits reduction of both natural and artificial samples through the formation of a surface layer which either traps air underneath or prevents a reducing atmosphere from penetrating the sample from the outside, or both. Reduction can only occur at the upper surface of this layer when it is intact, but removal of the layer allows reduction to proceed throughout the sample.

Buddington and Lindsley (1964) have shown that hematite will not reduce to magnetite at 500°C unless the oxygen fugacity is below 10^{-18} atmospheres (10^{-15} torr), and, if their curves can be extrapolated to lower temperatures, we must have an oxygen fugacity below 10^{-20} torr in order to reduce hematite at 300°C or below. The total vacuum achieved by our system is never better than 10^{-6} torr, so the buffering effect of hydrocarbons in the pump oils is sufficient to reduce the oxygen fugacity by more than ten orders of magnitude.

Ideally, we would carry out thermomagnetic analyses in atmospheres for which the oxygen fugacity could be measured or controlled. Unfortunately this is difficult because of the sluggish reactions of processes by which we measure or control atmospheres in much of the temperature range of interest to us. Oxygen electrodes do not work below 500°C. Also standard buffered reactions and gas mixing reactions (such as CO-CO_2) may not approach equilibrium at such low temperatures. One approach is to mix oxygen with an inert gas in the proper ratio, but it would be difficult to maintain a uniform mixture of a minute amount of oxygen in a large volume of inert gas. In any case, the appropriate fugacity could only be estimated since magnetite-hematite reactions probably do not approach equilibrium on the time scale of a typical thermomagnetic run at temperatures below 400°C.

Although we have shown that pump oils can be responsible for hematite reduction in thermomagnetic analyses under vacuum, reducing agents within the samples can also have an effect. We can not determine how important a role internal reducing agents played in the reduction of our or Schwarz's (1969) red samples. Perhaps they accounted for the greater amount of reduction of our natural samples, rather than the grain size effect discussed above.

In any case, other evidence indicates that natural red samples can contain enough reducing material that magnetite will be produced when the hematite is heated even if the atmosphere is not buffered by pump oils. Reduction often occurs when we carry out thermal demagnetization of red sediments at temperatures above 300°C. We heat up to 24 standard size cores in air in a cylindrical furnace 7.6 cm in diameter by 61 cm long. The free flow of air is somewhat restricted by loose fitting caps which cover the ends of the furnace. Reduction, when it occurs, is apparent from the loss of red color of the samples, which become darker and in some cases almost black. Another symptom is the occasional large increase of remanence during a thermal demagnetization step due to the aquisition of TRM by the magnetite in the magnetic field in the cooling chamber (less than 5 γ ordinarily) possibly augmented by fields of neighboring samples. Reduction is more commonly observed when a number of samples have been glued back together, but sometimes happens when none have been glued.

We have never observed hematite reduction during a thermomagnetic run in air, but such an example is seen in Fig. 2b of Kobayashi and Schwartz (1966). Their red sandstone sample shows the same increase of magnetization above 300°C and recovery of stronger intensity upon cooling that we observe for natural and artificial samples run in vacuum. They also show that samples from the contact zone of the same formation with an overlying basalt flow contain magnetite, indicating that heating in nature produced essentially the same results as heating in the laboratory.

7.   Paleomagnetic Implications

It is clear that reduction of hematite can be a very serious problem in thermal demagnetization of red sediments and paleointensity determinations on samples (including igneous rocks) containing hematite. Thermal demagnetization is generally favored over alternating field demagnetization for red samples (OULIAC, 1976; DUNLOP and STIRLING, 1977) so the reduction we often encountered may be a problem for many workers. There is not much chance to prevent reduction during heating. Even if a free flow of air is insured this will only inhibit reduction at the surface of the sample; the interior will reduce anyway. If the hematite and resulting magnetite are in the single-domain grain size range, the ratio of acquired magnetite TRM to original hematite CRM will be roughly equal to the fraction of hematite reduced times 250 (J_S (mag)/J_S (hem)) times ($H_{\text{cooling chamber}}/H_{\text{earth}}$). If the field in the cooling chamber is 10 γ and 10% of the hematite is reduced, the ratio is 0.5%, which is quite small. It will not be negligible, however, if a considerable amount of hematite remanence has already been demagnetized at lower temperatures, which is often the case. And increases in the ratio will result from increases in the amount of reduction and from furnace fields larger than 10 γ.

The parameter most easily subject to control is the field in the cooling chamber. We suggest the following precautions. 1) The cooling chamber field should be weak ($0.0 \pm 5\,\gamma$) and should be checked by a sensitive magnetometer before every run. 2) Thermal demagnetization at temperatures between 525°C and 575°C should be avoided if a separate cooling chamber is used because the batch would be transferred from furnace to cooling chamber in the blocking temperature range for single domain magnetite, and stray fields could produce a large PTRM. 3) Intensely magnetized samples should be separated from other samples. For example, a standard 2.5 cm core sample with an intensity of 10^{-3} emu/cc will produce a field of 5 γ at a distance of about 7 cm from its center.

Reduction of hematite during paleointensity determinations would be a more serious problem, because the large TRM acquisition by the newly formed magnetite would lead to anomalously low estimates of the paleofield. Careful control of the field would not be helpful in this case, since samples must be cooled in a field no matter which paleointensity method is used. Nor would it help to use hematite PTRM's acquired above the blocking temperature of magnetite (and nulling the field below 580°C). If some hematite has been reduced, it is not available to acquire a PTRM, so the paleofield estimate will still be affected. The ideal solution is to use samples which do not contain reducing agents and thus do not reduce. If samples which reduce must be used, it would be acceptable (but probably impractical) to study PTRM's lost and gained at temperatures below reduction. If the reduction is small, perhaps its magnitude could be estimated and a correction factor applied to PTRM's lost and gained in the region 580–675°C.

Suzanne Beske-Diehl assisted with magnetic measurements, Charles Keefer with X-ray diffraction analysis and Tim Drever provided clay samples. The work was supported by NSF grants GA-43388 and EAR75-14902.

REFERENCES

BRINDLEY, G.W. and R.F. YOUELL, Ferrous chamosite and ferric chamosite, *Mineral. Mag.*, **30**, 57–78, 1953.

BUDDINGTON, A.F. and D.H. LINDSLEY, Iron-titanium oxide minerals and synthetic equivalents, *J. Petrol.*, **5**, 310–357, 1964.

DOELL, R.R. and A. COX, Recording magnetic balance, in *Methods in Paleomagnetism*, edited by D.W. Collinson, K.M. Creer, and S.K. Runcorn, pp. 659, Elsevier, Amsterdam, 1967.

DUNLOP, D.J. and J.M. STIRLING, "Hard" viscous remanent magnetization (VRM) in fine-grained hematite, *Geophys. Res. Lett.*, **4**, 163–166, 1977.

ELDER, T., Particle-size effect in oxidation of natural magnetite, *J. Appl. Phys.*, **36**, 1012–1013, 1965.

GROMMÉ, C.S., T.L. WRIGHT, and D.L. PECK, Magnetic properties and oxidation of iron-titanium oxide minerals in Alae and Makaopuhi Lava Lakes, Hawaii, *J. Geophys. Res.*, **22**, 5277–5293, 1969.

JOHNSON, H.P., H. KINOSHITA, and R.T. MERRILL, Spinel formation by heating hematite in air and water-vapour, *Nature Phys. Sci.*, **239**, 151–152, 1972.

KERR, P.F., J.L. KULP, and P.K. HAMILTON, Differential thermal analysis of reference clay minerals, A.P.I. Research Project #49, Columbia University, 1949.

KOBAYASHI, K. and E.J. SCHWARZ, Magnetic properties of the contact zone between upper Triassic red beds and basalt in Connecticut, *J. Geophys. Res.*, **71**, 5357–5364, 1966.

MARSHALL, M. and A. COX, Magnetic changes in pillow basalt due to sea floor weathering, *J. Geophys. Res.*, **77**, 6459–6469, 1972.

OULIAC, N., Removal of secondary magnetization from natural remanent magnetization of sedimentary rocks: alternating field or thermal demagnetization technique ?, *Earth Planet. Sci. Lett.*, **29**, 65–70, 1976.

READMAN, P.W. and W. O'REILLY, Magnetic properties of oxidized (cation-deficient) titanomagnetites, *J. Geomag. Geoelectr.*, **24**, 69–90, 1972.

SAMARA, G.A. and A.A. GIARDINI, Effect of pressure on the Néel temperature of magnetite, *Phys. Rev.*, **186**, 577–588, 1969.

SCHWARZ, E.J., A recording thermomagnetic balance, *Geol. Surv. Can. Pap.*, **68-37**, 1–11, 1968.

SCHWARZ, E.J., Thermomagnetic analysis of some red beds, *Earth Planet. Sci. Lett.*, **5**, 333–338, 1969.

SHIVE, P.N. and R.F. BUTLER, Stresses and magnetostrictive effects of lamellae in the titanomagnetite and ilmenohematite series, *J. Geomag. Geoelectr.*, **21**, 781–796, 1969.

SHIVE, P.N. and J.F. DIEHL, Thermomagnetic analysis of natural and synthetic hematite, *Geophys. Res. Lett.*, **4**, 159–162, 1977.

Note added in proof

Recently we have observed reduction to magnetite during thermomagnetic runs in air. The sample were oil bearing sands and shales from the Alaskan north slope. There is some indication that further reduction to pure iron may have occurred as well, but our balance could not follow this process above 800°C.

Adv. Earth Planet. Sci., **1**, 123–145, 1977

Characteristics of First Order Shock Induced Magnetic Transitions in Iron and Discrimination from TRM

Peter WASILEWSKI

Astrochemistry Branch, NASA/Goddard Space Flight Center,
Laboratory for Extraterrestrial Physics,
Greenbelt, Maryland, U.S.A.

(Received July 21, 1977)

The probable magnetic history of iron and iron-nickel can be specified if the material in question can be resolved using reflected light microscopy. Microstructures imparted by shock transitions are distinct from those imparted by a thermal transition—*always*. These results are briefly summarized.

Fine particle iron ($<5,000$ Å) can exist in face centered cubic (fcc) and body centered cubic (bcc) states at ambient temperature. When ferromagnetic bcc iron is shocked above 130 kb (the pressure transition-P_T) a first order magnetic phase change—to antiferromagnetic ε phase (hexagonal close packed)—results in a demagnetization. Reversal to ferromagnetic bcc remagnetizes the specimen if an external field is present. Iron can also exist in a metastable fcc state at ambient temperature in the form of spherical precipitates elastically constrained in an fcc copper matrix. Plastic deformation associated with shock impact at levels up to 50 kb, which ensures sufficient plastic strain, but negligible shock heating, transforms the antiferromagnetic fcc iron to ferromagnetic bcc iron. These samples are being used to calibrate the shock and thermal mechanisms of magnetization in fine particle iron (size range for present data $\sim$250 Å–$\sim$1,200 Å). Experiments in axial fields demonstrates that the transformation takes place on a millisecond time scale within the designated field, as the sense of the shock induced axial vector is the same as the field sense. In fields as low as $10\,\gamma$ the axial component is oriented in the same sense as the shock normal. The amount of fcc iron transformed depends on both precipitate size and the peak pressures. The observed deformation induced anisotropy is dependent on particle size and shock level, but the 400–600 Å (B3, A3) set of samples shows the maximum degree of anisotropy at all levels of shock. Different sizes of precipitates appear to support different degrees of magnetic hardening. Anisotropy in magnetic hysteresis, using hysteresis ratios R_I (ratio of saturation remanence to saturation magnetization) and R_H (ratio of remanent coercive force to coercive force) indicates a shock induced anisotropy in remanence. The ratio of remanence after the shock to the saturation remanence is always less than the ratio of TRM to SIRM, but the shock induced remanence is more stable to demagnetization.

1.　Introduction

Solid solar system bodies—the terrestrial planets, moons, asteroids, and analyzed meteorites and lunar samples exhibit evidence of impact deformation i.e., high strain rate metallographic evidence in meteorites, the cratering record for Mars, the moon, Mercury, Phobos and Deimos, and the Earth, and the evidence from polarization curves for particulate asteroidal surfaces. Experimental evidence demonstrates that iron and the iron-nickel alloys are responsible for the magnetization of lunar samples (FULLER, 1974, 1975) and most meteorites (WASILEWSKI, 1975). Studies of lunar samples and meteoritic remanent magnetization together with measured lunar fields (SHARP *et al.*, 1973) and the Mercurian field (NESS *et al.*, 1975) argue emphatically for sample magnetic remanence and magnetic fields outside our terrestrial experience. In general terrestrial materials are magnetic by virtue of the iron oxides they contain, acquiring their remanent magnetization via thermal or thermochemical mechanisms in a dipole field of apparent regularity throughout geologic time. Alteration of the magnetic mineralogy is dominated by thermochemical processes, notably oxidation. In extraterrestrial materials, the iron-nickel alloys constitute the dominant magnetic mineralogy, being magnetized in fields of unknown intensity and configuration. If a change in oxidation state takes place, reduction may dominate except possibly for Mars. FULLER (1974) has reviewed the various processes and models invoked to explain the measured remanent magnetization in lunar samples. CISOWSKI *et al.* (1974, 1975) have obtained interesting results which support shock impact as a possible important factor in the magnetization of lunar samples, and in modification of the lunar crust magnetization configuration. Recent progress has been made in classifying the various thermal and dynamic remanent magnetization mechanisms for iron-nickel alloys and in demonstrating the significance of alloy composition and pre-shock and post-shock thermal history (WASILEWSKI 1973, 1974a, 1974b, 1975, 1976). In view of the overwhelming evidence for impact effects, it should be obvious that the competitive roles of dynamic and thermal remanent magnetization mechanisms be evaluated with iron-nickel alloys. The iron-nickel alloys, unlike spinels or other magnetic oxides, retain a microstructural record of high strain rate, diffusional, or transformation effects as described by WASILEWSKI (1974a). In iron, initial magnetization, demagnetization and remagnetization can take place at ambient temperature even though the Curie point is 780°C. It is therefore important to understand: (a) the efficiency of remanence acquisition for first order shock transitions compared to the second order Curie transition in identical fields, (b) how to separate shock induced remanence from thermoremanence, and (c) how to recognize remanence associated with partial recrystallization from remanence associated with various rapid transition type mechanisms.

2.   Mechanisms of Magnetization

Iron is conventionally ferromagnetic below ~780°C, the Curie point, where thermoremanence is acquired on cooling. It is possible to produce iron at any ambient temperature by various processes such as for example ion bombardment. In the extreme αFe_2O_3 can thus be converted to metallic iron (see YIN *et al.*, 1972, 1975). Table 1 lists the magnetic properties before and after ion bombardment, of a thin layer of αFe_2O_3 coating a non-magnetic high purity aluminum foil. The ratio of remanence associated with the reduced (Fe identified by electron spectroscopy by Dr. Yin) αFe_2O_3, to saturation remanence has the same value as the TRM/SIRM value for 1.5 μm Fe_3O_4 (TRM/SIRM$=0.018$) (PARRY, 1965). Similar magnetic changes can be expected, with samples containing iron at their surfaces when subject to ion bombardment and UV irradiation (HUGUENIN, 1974). Such surface specific processes would not be expected to contribute much to the overall magnetization change of a rock, but would be important perhaps in soils.

Table 1. Results of argon ion bombardment-reduction experiments.
Iron was identified by electron spectroscopy by Dr. Yin.

Magnetic remanence (RM)	
$<10^{-8}$ emu	Aluminum foil
(a) 0.006×10^{-4} emu	αFe_2O_3 on Al foil
(b) 0.007×10^{-4} emu	
0.331×10^{-4} emu (RM)	reduced foil (a)
16.77×10^{-4} emu (SIRM)	saturation remanence in 8.4 k Oe field

$$\frac{RM}{SIRM}=0.019.$$

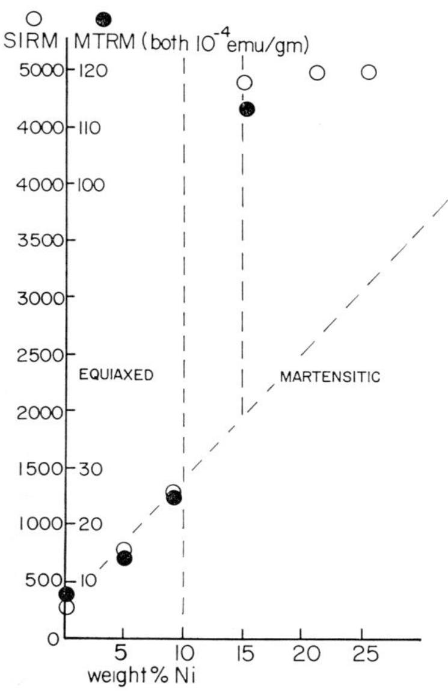

Fig. 1. Plot of (SIRM) saturation isothermal remanence (in 8.4 k Oe field) and (MTRM) martensitic remanent magnetization vs wt % Ni in FeNi spheres. The spheres were free fall solidified and cooled in the geomagnetic field. Each point represents a mean value for several spheres. The microstructure for Ni < 10 wt % is equiaxed ferrite, between 10 and 15 wt % a transition structure exists, and for Ni ≥ 15 wt % martensite is present.

Shock impact can produce initial magnetization in materials of specific compositions or those which have special crystallographic constraints. Complete or partial demagnetization and remagnetization following transitions induced by shock are also possibilities. Therefore, it is necessary to have discrimination criteria available when studying natural samples in order to identify which thermal or dynamic effects are responsible for observed natural remanent magnetization. Unlike the iron oxides, the FeNi alloys retain in their microstructure a vivid record of past thermophysical events. This feature enables us to immediately become quite specific about the magnetic history of a natural sample if we know what magnetic effects are associated with specific microstructural characteristics. The mechanisms of magnetization in FeNi alloys and their recognition criteria have been described by WASILEWSKI (1974a).

If a solid particle of FeNi is chemically homogeneous and cools rapidly enough to discount any diffusional phase separation, then martensitic remanence (MTRM) is acquired on rapid cooling to ambient temperature. Characteristic microstructure and magnetic properties are produced depending on the Ni content. A summary of

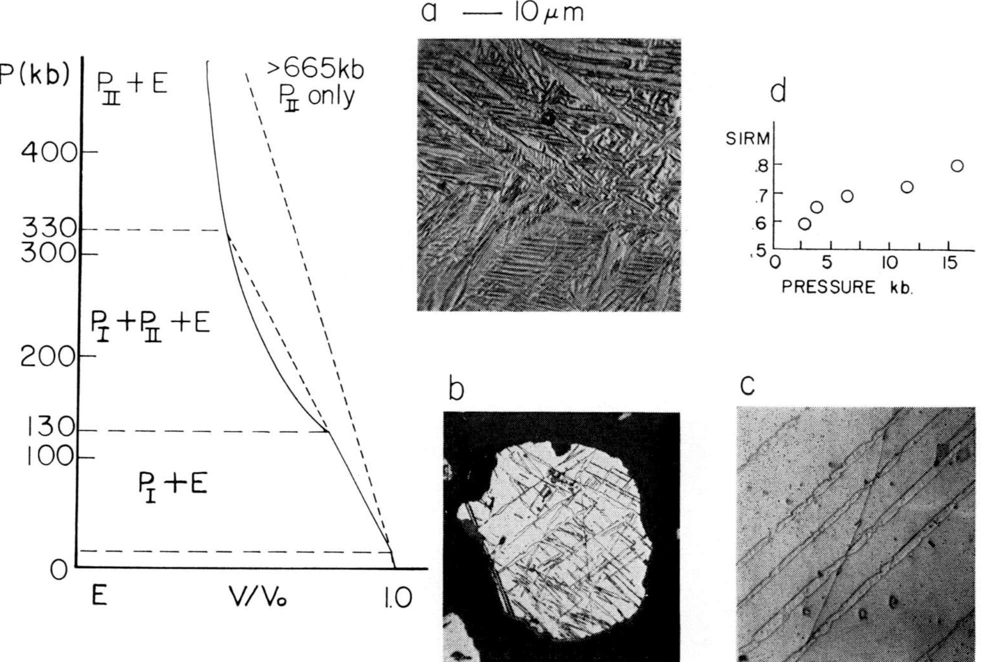

Fig. 2. Dynamic compressibility curve (schematic) for iron illustrating the wave structure in each zone P_I=plastic I wave, P_{II} plastic II wave, E=elastic wave. The pressure transition (P_T) at 130 kb is indicated. (a) micrograph of area of Odessa meteorite shocked to 200 kb. This is the transformation microstructure—a result of the bcc→hcp→bcc′ transformation. (b) twinned grain in Dalgety Downs chondrite. (c) twinned area in Campo del Cielo meteorite. (d) SIRM (saturation isothermal remanence) vs shock level in prior annealed Ferrovac E iron.

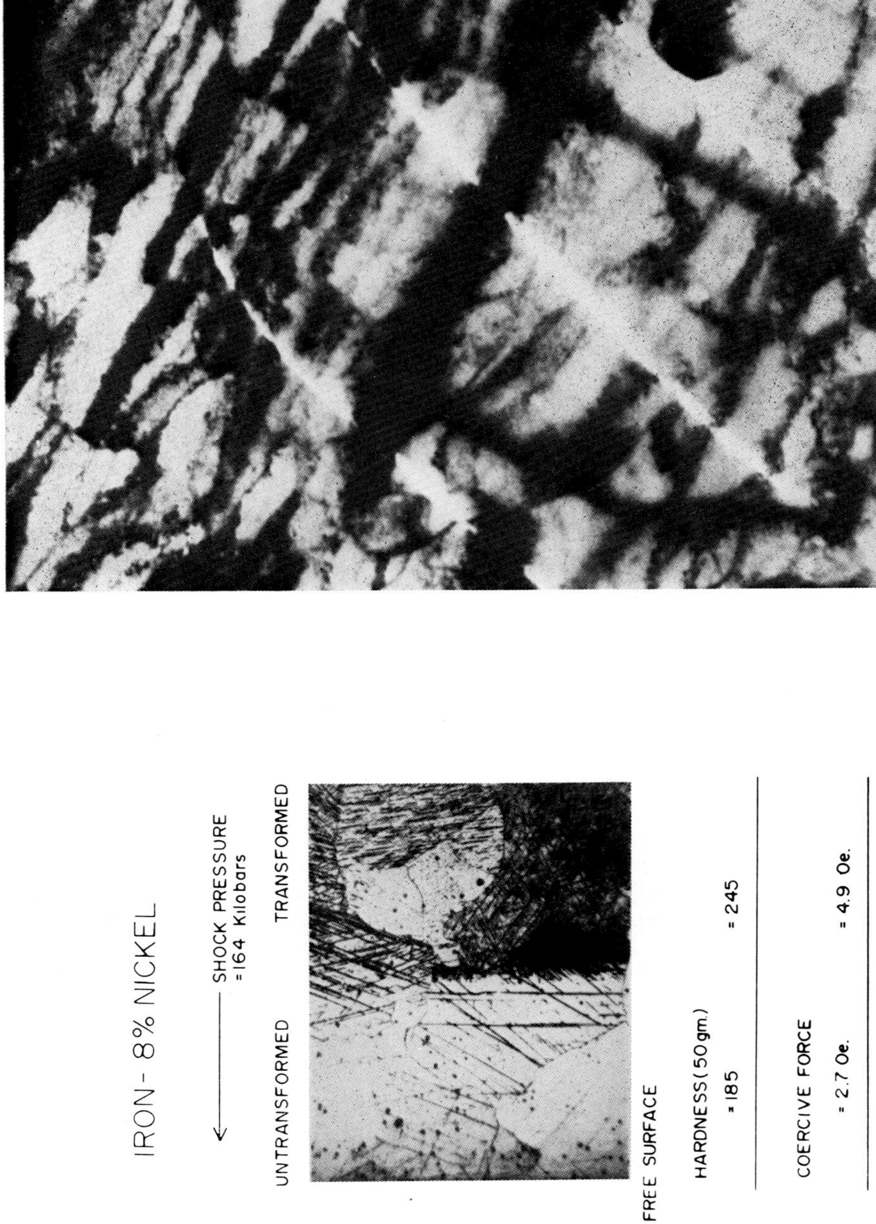

Fig. 3. Optical micrograph of Fe-8% Ni alloy shock-loaded to 164 kb showing transformed and untransformed regions. Hardness data are from LESLIE et al. (1965) and coercivity data are results of the present investigation. High voltage Lorentz micrograph of shock-loaded Fe-8% Ni taken close to focus to illustrate both domain boundaries and heavy dislocation substructure. Note that domain structure is much coarser than scale of defect distribution. (Approx. ×15,000).

a — 0.01 mm
304 SS
—IMPACT SURFACE
Fe 30 Ni
b — 0.01 mm
c — 10 μm
d — 10 μm
e — 10 μm
f — 10 μm

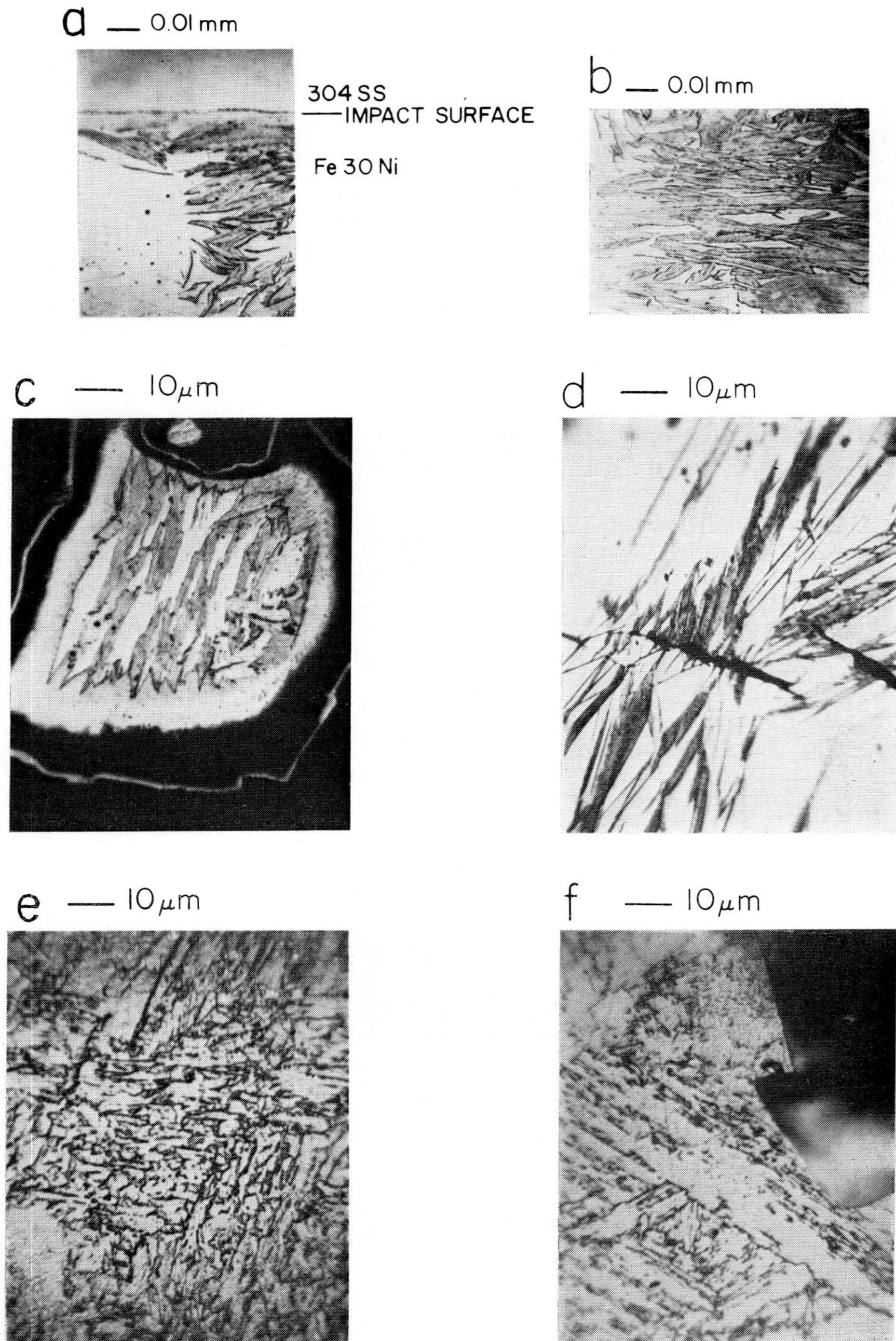

Fig. 4

the results for free fall solidified spheres, cooling in the geomagnetic field, is presented in Fig. 1 (from WASILEWSKI and LARSON, 1977). For spherical specimens studied, the mean MTRM value for the Fe composition is 7.55×10^{-4} emu/g and the SIRM value is 282.7×10^{-4} emu/g. For low Ni alloys ($<10\%$) the Ni contribution to MTRM is $\sim 2 \times 10^{-4}$ emu/g per wt % Ni and to SIRM is $\sim 125 \times 10^{-4}$ emu/g per wt % Ni. The total remanent moment (M) is made up of the base iron contribution (M_{Fe}), the nickel contribution (M_{Ni}), and the substructure contribution (M_{ST}) i.e., $M = M_{Fe} = M_{Ni} + M_{ST}$ (see WASILEWSKI and LARSON, 1977).

The response of a material to pressure can be described by its dynamic compressibility curve (Fig. 2). As seen in the figure a discontinuity exists at 130 kb. Above 130 kb a magnetic transition is recorded (FOWLER *et al.*, 1961, SMITH, 1958 and others), and below the transition twinning is the dominant deformation mode (SMITH, 1958 and others). The wave structure for various pressure regimes is indicated, and significantly when pressures exceed ~ 350 kb shock heating becomes important and is responsible for modification of magnetic properties (WASILEWSKI, 1976). Micrograph (a) characterizes the transformation microstructure in a sample of Odessa subjected to $P = 200$ kb. This transformation microstructure is evidence of the bcc$\rightarrow$hcp$\rightarrow$bcc$'$ reversible magnetic transition. Micrographs (b) and (c) show the microscopic appearance of shock induced twins in a grain from the Dalgety Downs chondrite and the Campo del Cielo meteorite respectively.

Evidence for the role of twinning is shown in Fig. 2d. JOHNSON and ROHDE (1971) have shown that the twin density is a monotonic function of the loading level in Ferrovac E iron shocked between 0 and 15.5 kb. The results in Fig. 2d (WASILEWSKI and DOAN, 1973) provide the correlation between twin density and the magnitude of SIRM.

When $P > P_T$, two plastic waves exist because of the transition; the faster moving transition wave reflects from the free surface as a rarefaction wave and moves back into the sample to collide with the slower but higher pressure wave. The collision front marks the position of advance of the high pressure wave which is more or less abruptly diminished. This situation is indicated in Fig. 3a (SMITH, 1958; FOWLER

Fig. 4. Microstructures associated with specific thermal and shock processes which identify the nature of the mechanism(s) of magnetization (a) 304SS-Fe30Ni interface. The antiferromagnetic 304SS has been partially transformed to a martensitic product, with anisotropy parallel to the shock front. Immediately below the impact surface the Fe30Ni is clear fcc. The clear area to the left in the photo is clear fcc running perpendicular to the interface. The martensite plates immediately below the clear zone are parallel to the shock front and as shown in (b) there is a tendency for the martensite to run parallel to the shock front. The Fe30Ni was quenched to liquid nitrogen temperature and held to transform the fcc to bcc martensite prior to the shock. X-ray analysis identified the dominant phase as fcc after the shock. Much of the bcc martensite has been shock transformed to fcc martensite. (c) This grain has been reheated following deformation (note lines which indicate faults) as evidenced by the concentricity of the etching zones. The martensite was formed on cooling following the reheating event (Alleghan chondrite). (d) Martensite in Odessa shock loaded to 200 kb. (e, f) Annealed martensite within grains in Renazzo chondrite and Lunar sample 14310 respectively.

et al., 1961). The untransformed zone ($P \leq 130$ kb) is twinned only, the transformed zone has experienced the bcc→hcp→bcc′ transition (WASILEWSKI, 1973).

The domain structure of shock hardened alloys is coarser than the scale of the defect distribution in alloys which have not been shocked to pressures large enough to result in significant shock heating (Fig. 3b). This describes why the H_c values even in transformed alloys, are only a factor of 2 or 3 larger than in twinned alloys. However the defect structure might provide pinning sites for stable remanence. An austenitic 304 stainless steel disc impacting an Fe 30 wt % Ni disc illustrates another important aspect of impact magnetization. First the impact produces magnetic anisotropy in both discs. The 304 SS is initially antiferromagnetic and fcc which explains its characteristic 'non magnetic' properties. Plastic deformation produces a martensitic transformation which results in ferromagnetic bcc partially aligned parallel to the shock front (unpublished research—see also ABUKU and CHIKAZUMI, 1967—case of conventional deformation) Fe30Ni which was quenched and held at 78°K, transformed to bcc martensite. The shock transformed most of the bcc martensite to fcc martensite and clear fcc. This is a magnetic transition of first order. The situation is indicated in Fig. 4a and 4b.

In meteorites thermal (Fig. 4c) and shock (Fig. 4d) martensite are commonly found. If martensite is annealed (as in Fig. 4e and 4f) by reheating, this can be recognized and thermal demagnetization experiments should therefore consider the existence of this overprint. In all of the cases presented in Fig. 4, the microstructure is diagnostic of the primary mechanism of magnetization, and subsequent thermal overprints on the primary microstructure are identifible. These microstructural considerations provide a strong argument against acceptance of any paleointensity data derived from an experimental program which has not considered such data in planning the program.

3. Fine Iron Precipates

Iron can exist in four distinct states within a copper disc (CAMPBELL and CLARK, 1974; WINDOW, 1972 and others).

—Iron in solution, the material is paramagnetic—Antiferromagnetic face centered cubic iron precipitates, diameters up to 2,000 Å with Neel points >30 K but <70 K—Clusters of iron atoms with sizes up to 25 Å—Ferromagnetic body centered cubic iron.

It is highly unlikely that FeCu alloys exist in natural samples. The dilute iron (1.5 wt %) in copper alloys have been chosen for shock calibration and strong evidence will be presented for utilization of the alloy to understand all aspects of magnetic remanence in fine particle iron for the following reasons:

1) The concentration of iron, hence the precipitate size and mean particle spacing can be arbitrarily specified and controlled.

2) By using an appropriate precipitation anneal time the size of the precipitate can be varied according to need for specific experiments.

3) Magnetic shape effects are well controlled as the precipitates remain spherical up to $\sim 1,000$ Å and become cubic in habit thereafter until extreme overaging produces a Widmanstatten pattern of Fe in Cu.

4) As will be described in a latter section, the sense of the external field is recorded in the remanent magnetic state associated with the transformation, therefore a second shock with the same sample in a field different from that for the first shock provides an ideal experimental framework for evaluating the role of shock in modifying existing remanence, in quantitative terms.

5) The effects due to recrystallization of shock hardened fine particle iron, and a comparison of TRM, shock induced remanence, recrystallization remanence effects etc. can be compared in the same sample, without modification of shape, size or interparticle spacing of the transformed iron precipitates.

In the experiments, plastic deformation associated with shock wave impact transforms the antiferromagnetic fcc iron precipitates to ferromagnetic bcc at ambient temperature. Magnetic remanence is acquired on a millisecond time scale in either zero or any controlled external field. When an axial field of $H = \pm 0.5$ oersted is applied, the axial remanent vector has the same sense as the field, verifying that during rapid first order shock transitions in weak fields, the field sense is recorded in the transformed precipitates.

4. Experimental Technique

Alloys were prepared by induction melting of 99.999 % purity iron and copper after which the alloys were homogenized at 1,025°C for 4 hours in a vacuum furnace. After the 0.500 inch diameter by 0.250 or 0.125 inch thick discs were machined, they then were precipitation annealed at 650°C or 750°C for various times to produce the face centered cubic iron particles. The gas gun used in the shock loading experiments was constructed by Dr. Frank Rose of the Naval Surface Weapons Center. The barrel was made of copper, all gun supports and the sample supports were constructed of wood, and the velocity measuring was done with 'light pipes', all in the interest of minimizing spurious magnetic fields. The sample itself was contained in a copper sample assembly and the projectile consisted of a copper disc mounted on one end of a thin walled hollow nylon cylinder. The projectile weighed 10.0 grams. Calibration curves were constructed and the measured velocity for each experiment was used to evaluate the peak impact pressure. The sample assembly was sited in an open ended mu metal shield which contained a solenoid used to produce axial fields. Soft recovery was achieved by catching the copper sample assembly in a rag box backed by plastic foam. In no case was the sample lost. Magnetic remanence measurements were recorded before and after the shock. Magnetic hysteresis results were obtained only on a single sample from each sample set, (for the non shocked case), but each shocked sample was measured. Magnetic hysteresis results were obtained with a Princeton Applied Research vibrating sample magnetometer (VSM). Vector measurements of residual magnetization were obtained with a Superconducting

Technology cryogenic magnetometer. AF demagnetization was achieved with a Schoenstedt demagnetizer.

5.   Experimental Results

5.1   Microstructure

Polycrystalline copper discs with 1.5 wt % iron were precipitation annealed for various times at 650°C and 750°C. Spherical particles of fcc iron were developed with the size dependent on the anneal time for the 1.5 wt % iron concentration (see Table 2). The particle sizes were estimated from electron micrographs. The coherency strains associated with the fcc precipitates are imaged as strain lobes which are due to the diffraction contrast in the strain field (PHILLIPS and LIVINGSTON, 1962). An example is shown in Fig. 5a. When the particles transform to bcc during the impact, the accommodation coherency strain, hence the lobe contrast disappears. The intersection of primary dislocations with the particle and tangled dislocation loops at the particle interface characterize the transformed precipitate which is now ferromagnetic (Fig. 5b). At higher magnifications the structure in the particle is observed as the light and dark banded diffraction contrast (Fig. 5c). Two interpretations exist for the light-dark contrast. EASTERLING et al. (1967, 1969) consider that the bands represent alternation of bcc martensite and retained fcc phase. KUBO et al. (1975) suggest that the single crystal fcc iron particle transforms to two twin related bcc martensites. An attempt was made to discriminate between the two competing explanations by cooling to 4 K (below the Neel point of fcc phase) in an external field of 10 k Oe to determine if exchange coupling could be detected. No

Table 2.   Description of the size and shape of the precipitated particles subject to specific precipitation anneals and the initial magnetic state for samples studied.

Samples	Growth stage	PPT anneal	Magnetic state
A1, B1	Cluster of atoms ($\leq$12 atoms). Some small fully coherent precipitates. ($<$100 Å)	None—these are the starting discs quenched from 1,000°C.	Very minor ferromagnetic impurity paramagnetic copper mainly (Fe in solution mainly).
B2, A2	Small fully coherent spheres. (200 Å–350 Å)	650°C–10 h	Minor ferromagnetic impurity—mainly antiferromagnetic fcc iron.
B3, A3	Fully coherent spheres. (400 Å–600 Å)	650°C–100 h	Antiferromagnetic fcc iron—minor ferromagnetic impurity.
B4	Fully coherent to semi-coherent fcc spheres and 'cubes'—some of the larger precipitates have transformed. (850 Å–1,100 Å)	750°C–24 h	Initial ferromagnetic component increases by factor of 4 over the (B2, A2) and (B3, A3) sample sets. Most iron is antiferromagnetic fcc.

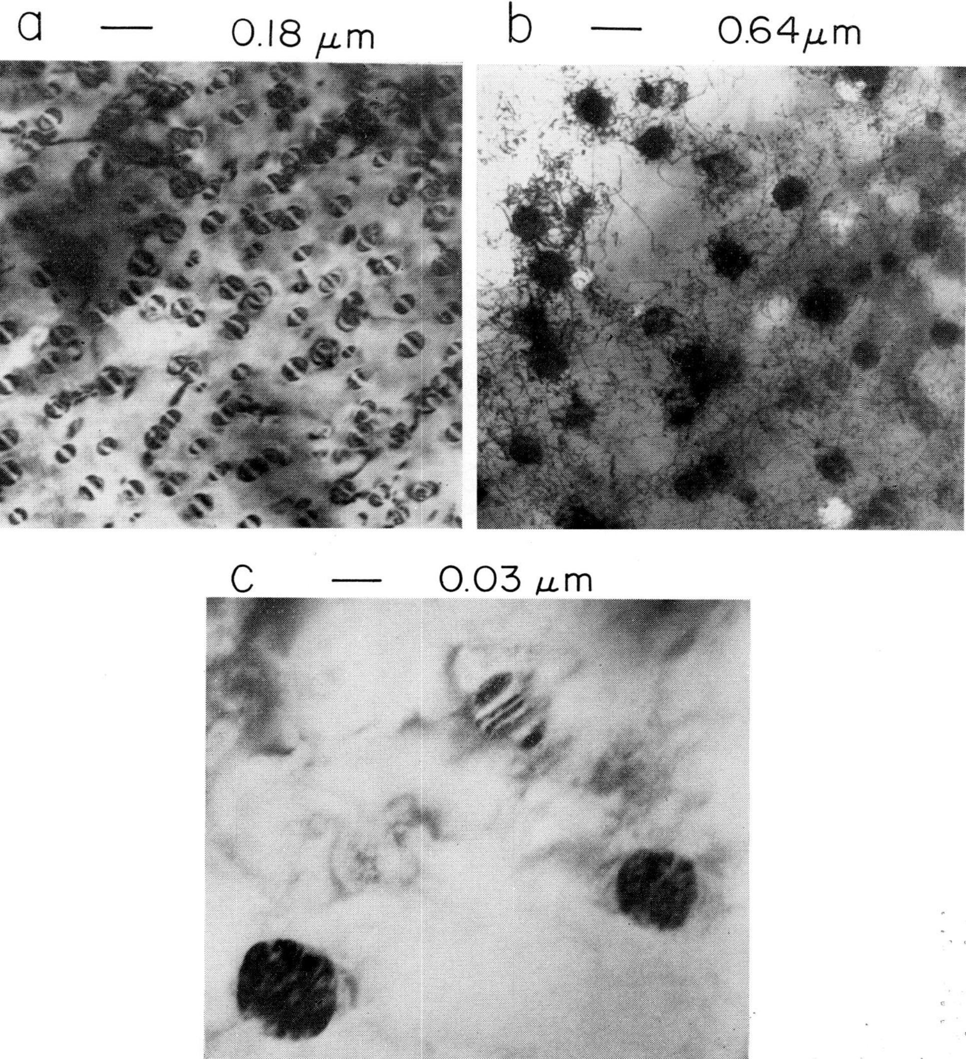

Fig. 5. Electron micrographs of iron precipitates in copper discs. (a) Characteristic strain lobe contrast associated with diffraction in the accommodation satrain field. The contrast effects make the particle appear larger than it really is (Sample B3). (b) This sample has been shocked at 25 kb resulting in ~90% transformation (sample SB4). The strain lobe contrast disappears, dislocations decorate the precipitate interface and matrix and as shown in (c) at higher magnification the precipitate particles contain alternating light-dark bands. This banded structure (see text) has been interpreted as bcc martensite + retained fcc or twinned martensite and appears unresolved at present.

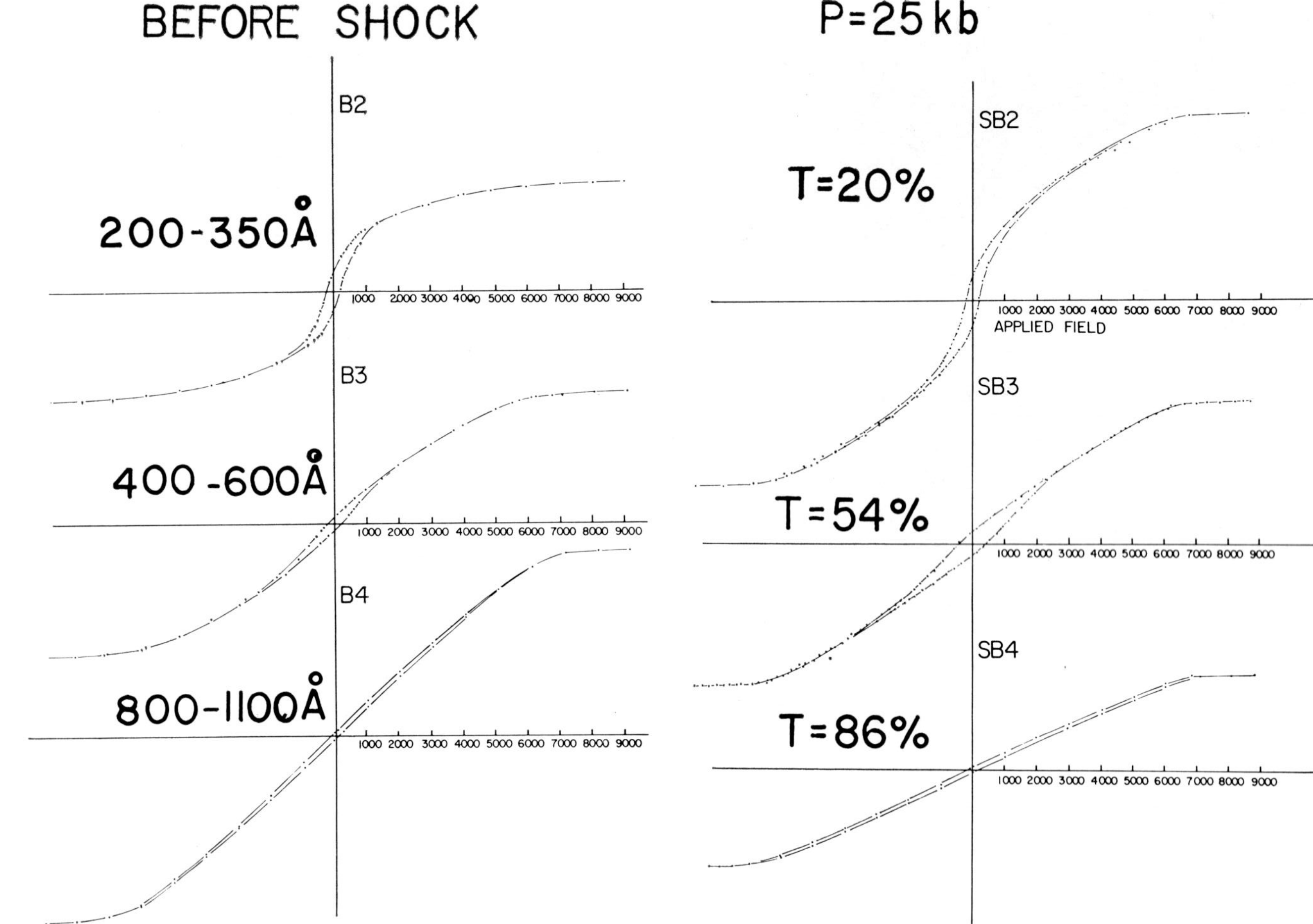

Fig. 6. Magnetic hysteresis loops for B2, B3 and B4 samples before and after a 25 kb shock. The range of observed particle diameters is indicated. T is the percentage of fcc transformed, based on saturation magnetization measurements.

loop shifting or otherwise anomalous hysteresis was observed in the preliminary experiments.

5.2 Magnetic hysteresis

It was determined that the solution annealed specimens (i.e., 1,025°C for 4 hours) were paramagnetic with a very minor ferromagnetic impurity. Precipitation annealed samples contained minor ferromagnetic impurities as shown in Fig. 6 (before shock). The impurity amounts to ~4% of possible iron in B2 and B3 and ~18% in B4. The % transformed indicated in Fig. 6 (after 25 kb shock) provides clear evidence that the transformation for a given plastic strain is precipitate size dependent. The indicated precipitate size was estimated from electron micrographs (see Table 2—refinement is in progress). All samples were prepared under identical vacuum annealing conditions, and it would be expected that if oxidation was in any way significant, the B4 alloy would be affected the most. This does not appear to be the case, as the ramp shape to the B4 loop before and after the shock, and the saturation field $H_S = NI_S$ is well defined. This consideration was necessary since each of loop shapes is distinct, but for particle sizes $\gtrsim 1,000$ Å (such as set B5 which is not considered in this paper—aged 750°C-100 hours) the loop shape is identical to B4. Therefore we must explain the loop shape in terms of the precipitate size, shape and interparticle spacing. We are evaluating the possible role of copper as a substitutional element in the Fe precipitate, which may be significant as the particle size decreases. Table 3 lists the magnetic hysteresis loop data for sample sets (B2, A2) and (B3, A3). For the (B2, A2) set with precipitate size between 200 Å and 350 Å little hysteresis loop

Table 3. Magnetic hysteresis loop data for the (B2, A2) and (B3, A3) sample sets.

Sample	P(kb)	$\varepsilon_x = \dfrac{\Delta v^{1)}}{v_0}$	$T\%^{2)}$	H_C(Oe)	H_R(Oe)	R_H	R_I
B2	0	0	0	175	430	2.46	0.23
SB2D	10	0.0066	4%	170	400	2.35	0.15
				180	350	1.94	0.17
SB2	25	0.017	20%	190	365	1.92	0.15
				145	300	2.07	0.14
SB2	40	0.0272	38%	170	310	1.82	0.22
				—	—	—	—
SA21	50	0.0338	34%	170	310	1.63	0.25
				—	—	—	—
B3	0	0	0	200	1,200	6.0	0.045
				210	1,250	5.95	
SB3	25	0.017	54%	395	1,250	3.16	0.073
				175	920	5.25	0.029
SA31	50	0.0338	100%	580	1,600	2.71	0.095
				120	1,280	10.66	0.027

H_C, coercive force; H_R, remanent coercive force; R_H, ratio H_R/H_C; R_I, ratio of saturation remanence to saturation magnetization; P, shock pressure in kilobars; $\varepsilon_x = \Delta v/v_0$, uniaxial strain; $T\%$, percentage of fcc→bcc transformation.

1) Uniaxial Strain—$\varepsilon_x = \Delta v/v_0$ where v is specific volume.

2) $T\%$, percentage of total iron precipitates transformed from fcc→bcc.

anisotropy is noted for measurements parallel and perpendicular to the impact direction. On the other hand, the (A3, B3) set is remarkably anisotropic. These results argue strongly for a particle size dependence of deformation induced anisotropy.

Notable in the hysteresis loops is the behavior between $\pm H_R$ (H_R is the remanent coercive force). In the two examples given (Fig. 7), the arrow indicates reduction of field from 12.5 K Oe to the point B, at $H=0$, which is the saturation remanence. Field reversal to point A defines the remanent coercive force (H_R) which is the back

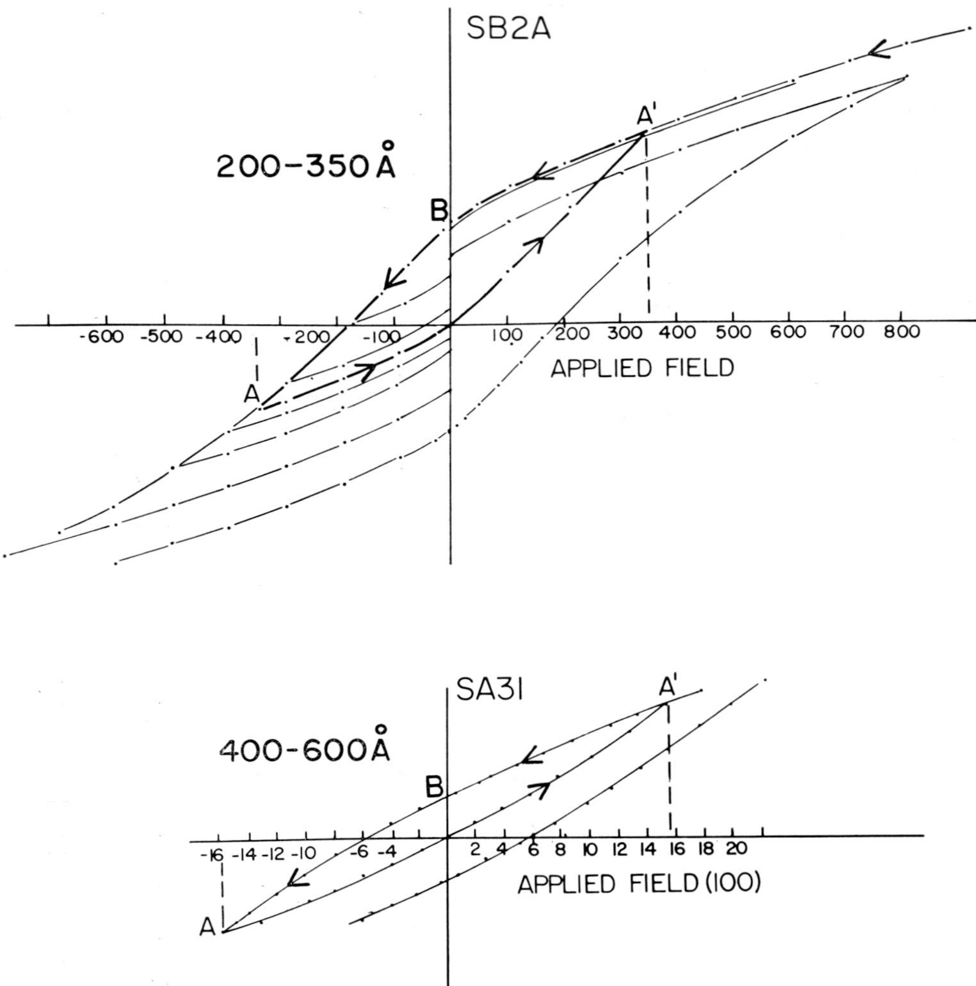

Fig. 7. Enlarged region of hysteresis loops for SB2A and SA31 samples measured in direction parallel to shock axis. Arrow indicates return from saturation at a maximum field of 12.5 k Oe. The point B is the saturation remanence. The back field needed to reduce remanence to zero is defined at point A. Application of a positive field equivalent to H_R and then reduction of the field to zero results in recovery of the saturation remanence. The path is indicated with arrows.

field needed to reduce remanence to zero (point 0). On taking the field to a value equivalent to H_R in the positive direction (point A') and then reducing the field to zero, saturation remanence is recovered. Additional experiments are needed to develop a detailed explanation for this behavior. The same behavior was noted in SB3 (see Fig. 6—after shock) and in SA31. All the magnetic hysteresis occurs in the range $\pm 2H_R$. (This value is $\sim 2,500$ Oe for SB3 shocked to 25 kb and $\sim 3,200$ Oe for SA31 shocked to 50 kb.) The rest of the loop, i.e., between $2H_R$ and the saturation field $H_S = NI_S$ appears to be due to average demagnetization characteristics for the transformed precipitates in the copper polycrystal.

5.3 Magnetic remanence

Samples were shocked in three field configurations: $H(+)$ in the direction of the shock normal, $H(-)$ in the opposite direction, and $H=0$ ($\pm 10\gamma$) according to measurements with a fluxgate magnetometer. Some of the axial vector measurements are presented in Table 4. All sample measurements labelled (before shock) refer to the thermoremanent components acquired on disc preparation. The important record in the table is that in an axial field, $H(+$ or $-)$ the axial component always has the same sign as the field after transformation, and for the case $H=0$, the axial component is in the direction of the shock normal. The transverse components, in some cases of the same magnitude as the axial components, are considered due to the polycrystalline nature of the discs where the {111} directions are oriented at different angles to the shock normal. This will be considered in the discussion.

Demagnetization characteristics are summarized in Fig. 8. The non-shocked sample (B3B) which contains only the thermoremanent component is least stable to demagnetization. A small amount of precipitates will always be transformed due to grain boundary locations, precipitation near faults in the crystallites, etc. This feature provides us with a comparison of the TRM/SIRM ratios for highly dilute dispersion of iron. In a set of experiments designed to determine the effect of specimen thickness on attenuation of the shock, the single 0.250 inch thick disc was replaced by two 0.125 inch thick discs in the configuration shown in the lower left of Fig. 8. The dimensional changes of each disc in the stack (A33 lead-A34 follow)

Table 4. Axial magnetic vector data for samples shocked in various configurations[1].

Configuration	Sample	Pressure	M_Z(before shock) $\times 10^{-4}$ emu	M_Z(after shock) $\times 10^{-4}$ emu
	SA21	$P=50$ kb	-17.20	$+\ 4.02$
$H\rightarrow(+)$	SA22	$P=50$ kb	$-\ 1.54$	$+11.70$
$P\rightarrow$	SA31	$P=50$ kb	$-\ 3.45$	$+\ 3.87$
$H\leftarrow(-)$	SA24	$P=50$ kb	$-\ 3.65$	-10.66
$P\rightarrow$	SA33	$P=50$ kb	$-\ 3.58$	$-\ 7.40$
$''H=0''$	SB3C	$P=50$ kb	$+\ 0.693$	$+\ 2.12$
$P\rightarrow$	SB4B	$P=25$ kb	$+\ 2.700$	$+\ 6.232$

[1] $H(+)$, axial field in direction of shock normal.

  $H(-)$, axial field in direction opposite shock normal.

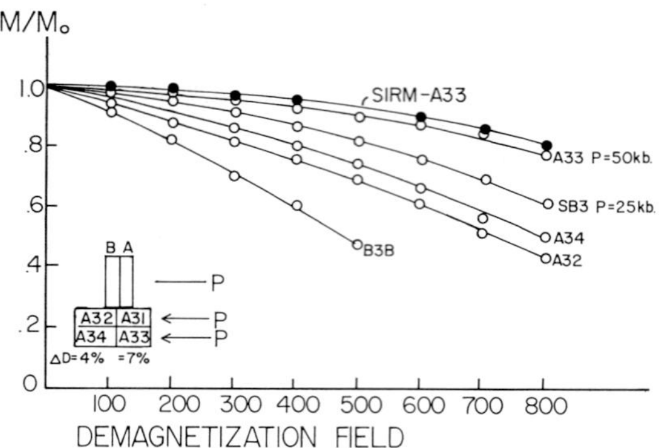

Fig. 8. Demagnetization curves for unshocked B3B containing thermo remanence, and samples which have been shocked. Two sample configurations have been used in the experiments; SB3, a 0.250 inch thick disc was shocked at 25 kb. Stacked samples of lead disc and follow disc, as shown, were shocked to evaluate shock structure and attenuation of the input shock level. A33 is the lead disc and A32 and A34 are follow disc curves. The lead disc A33 suffered a 7% change in thickness and the follow disc A34 suffered a 4% change in thickness verifying that plastic deformation attenuates with thickness. The lead disc also has an increased stability to demagnetization. The SIRM demagnetization curves for A33 is essentially similar to the shock remanence demagnetization curve.

are indicated. The lead disc experienced substantially more plastic deformation than the follow disc. Samples A32 and A34 which were follow discs are less stable to demagnetization than the lead disc A33. The two disc experiments had a peak input shock of 50 kb. The SB3 disc which was a 0.250 inch thick disc had a peak input shock of 25 kb. The results indicate that magnetic stability increases with the amount of shock deformation. Note that the SIRM demagnetization curve for sample A33 is essentially identical with the demagnetization curve for remanence acquired on shock transformation.

The efficiency of remanence acquisition is here defined as the ratio of remanence due to a specific mechanism in a weak field to the saturation remanence (Table 5). The samples A1, 003, and 002 contain TRM since they were quenched in water after solution anneal at 1,025°C. The TRM/SIRM values range from 0.013 to 0.04, which is similar to the metal spheres 0.023 to $\sim$0.057. The iron deposit produced by sputtering reduction of α Fe_2O_3 has an RM/SIRM value of 0.019 and the TRM/SIRM ratio for 1.5 μm Fe_3O_4 is 0.018. The shocked samples are quite distinct having ratios of 0.002 to 0.006. Samples B1 and 001 are samples which were subjected only to the solution anneal and hence we are looking at the remanence of the minor ferromagnetic impurity. Samples A33 and A34 (see Fig. 8) were shocked together in the A-B stack configuration, and the 0.0006, and 0.002 ratio values would suggest that

Table 5.   Values for the remanence ratios.

Sample	TRM/SIRM	SRM/SIRM
A1	0.013	—
003	0.039	—
002	0.043	—
Shocked	RM	
001	before 2.3 $\times 10^{-4}$ emu	
	after 0.102×10^{-4} emu	0.0079
B1	before 0.532×10^{-4} emu	
	after 0.134×10^{-4} emu	0.0058
A34		0.002
A33		0.0006

Reduced iron by ion bombardment RM/SIRM=0.019 (see Table 1)
(PARRY, 1965) 1.5 μm Fe_3O_4=TMR/SIRM=0.018

Metal spheres[1]	MTRM/SIRM	
Fe-A	0.057	—
Fe-C	0.023	—
Fe5Ni-B	0.026	—
Fe15Ni-B	0.035	—

Titano magnetites-LEWIS (1968) (0.007–0.014)

TRM, thermo-remanent magnetization; SIRM, saturation isothermal remanence; SRM, remanence after shock; MTRM, martensitic remanence.
[1]   Free fall solidification and cooling in geomagnetic field.

the amount of plastic deformation is important for any particle size, not just multi-domain particles. Even if a sample undergoes a shock transformation, the remanence level is at least one order of magnitude lower than that expected for a TRM, for example. The shock impact produces more ferromagnetic material which is more stable to demagnetization and has a definite magnetic anisotropy, but the ratio of the observed remanence to the SIRM is significantly lower than the TRM/SIRM ratio.

6.   Discussion

Spherical iron precipitates are formed during aging of dilute Cu—1.5 wt % Fe discs. The precipitates are face centered cubic (fcc) and antiferromagnetic at ambient temperature and will remain so until subjected to plastic deformation. The fcc precipitates are preserved because of the elastic constraints of the fcc copper matrix ($\sim$1.5 % mismatch at most). The accommodation elastic strains are imaged as lobes in thin foil electron microscopy examination (see Fig. 5a). Shock experiments consisted of impacting a thin copper disc on the alloy sample. In estimating strain levels and general dynamic response, the dynamic compressibility of copper and the dilute alloy were considered identical. At all levels up to 50 kb the plastic deformation was sufficient to activate the transformation and did not produce any thermal rise unless the impact was grossly nonplanar,—the extreme being a weld of samples and projectile. The experiment then can be described as a non-magnetic→ magnetic rapid first order transition in controlled external fields.

Shocked specimens were polycrystal discs which had been precipitation annealed after being machined into the disc shape. Newkirk (1957) performed a series of experiments intended to shed light on the mechanism of precipitation. He noted that equiaxed iron pattern develops when the alloy is slowly cooled from the solution temperature to the aging temperature and then aged, whereas, quenching to room temperature and then reheating to the aging temperature produces a {111} Widmanstatten pattern. He interpreted this to mean that active nucleation develops on the primary slip system in copper. Examination of diffraction patterns led Newkirk to conclude that the fcc iron pattern and fcc copper pattern are identically oriented. Slip is the basic deformation mode in copper; therefore {111} slip planes in the fcc iron precipitate coincide with the {111} orientation in the copper matrix and respond anisotropically to the slip system when deformed. On a crystallite basis therefore an orientation dependence exists. Pure copper cold rolled in one direction produces a strong texture such that crystallites tend to become preferentially oriented with respect to the rolling direction (see Williamson et al., 1976).

Livingston and Becker (1958) suggest that both Co-Cu and Fe-Cu alloys owe their deformation induced anisotropy and Hc increase to ΔN, i.e. shape change. The anisotropy can be calculated from $K=(\Delta N/2)I_s^2$ which is valid for homogeneous compression. However, agreement between theory and experiment is reasonable only for small compressive strain. Disagreement is also greater for increasing initial precipitate size. Results in this paper document a particle size dependence for induced anisotropy. Since copper is soft and iron is a hard particle (Fe is elastically harder than Cu by a factor of 1.7), we might expect some particle size control as Woolhouse (1974) and Matsuura et al. (1973) have indicated. It is plausible to include a significant shear component in the deformation process. In addition, the transformation is martensitic. Electron microscopy confirms the presence of banding in the transformed precipitates, most easily observed in the B4 samples (≥ 800 Å, see Fig. 5c). Matsuura et al. (1973) have shown that in conventional modes of deformation the lower the temperature and the larger the particle size the more rapidly the transformation occurred.

Calculations (Easterling and Weatherly, 1969) have been made for a critical precipitate size, below which the martensite transformation cannot be induced. This size is ~200 Å which is approximately the critical single domain size. The likely shape for a martensite product in a fine particle is a platelet; the needle is a poor shape for accommodating a pure shear transformation. Therefore, it appears that shock deformation induced magnetic anisotropy and resultant high magnetic stability is size dependent. This would apply to the present experiments or for fine particles in any deformed matrix. It is reasonable to assume that a 500 Å Fe sphere which has been structurally hardened has a higher magnetic stability than a 250 Å particle which has not, simply because of microstructural anisotropy. Further experimental studies should shed light on this contention.

KUBO *et al.* (1975) suggest that single crystal fcc iron transforms to two twin related bcc martensites. However, such internal twins are not found in bulk Fe martensite and only dislocation lines have been observed in lath martensite. Because fcc iron precipitates have transformed at ambient temperature Kubo *et al.* invoke a similarity argument to suggest the transformation temperature favors twinning. This question deserves further consideration. The fact that no exchange anistropy was observed in the 200 Å–350 Å samples might suggest that only in the larger

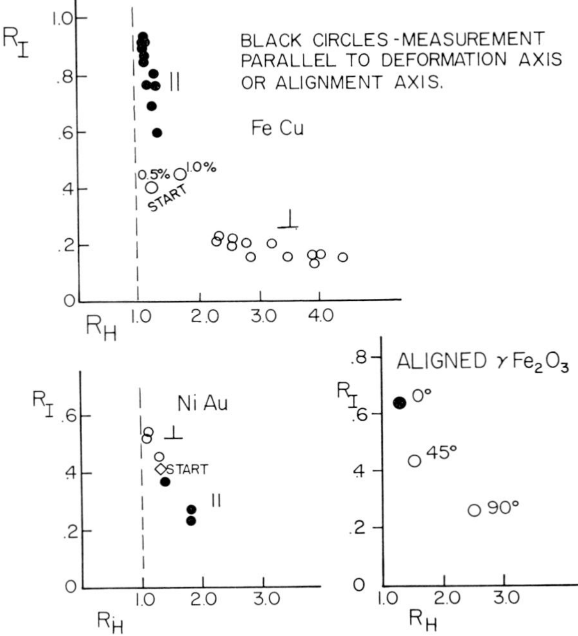

Fig. 9. R_I (ratio of saturation remanence to saturation magnetizations) and R_H (ratio of remanent coercive force to coercive force) are hysteresis loop ratios. (a) FeCu wires were pulled by LOTHIAN *et al.* (1958). The isotropic R_I and R_H values at the start of the experiment are indicated for a 0.5 wt % and 1.0 wt % alloy. Their data has been replotted to demonstrate the increasing shape anisotropy induced by pulling. The maximum R_I values and R_H are for the cases parallel and perpendicular respectively to the axis of wire elongation, at the last step of pulling. (b) The Ni-Au wires exhibit completely opposite behavior. The maximum R_I and R_H values are measured after the initial wire pull and occur for perpendicular and parallel measurements compared to the opposite in FeCu. This is due to Ni having large negative magnetostriction. The values of R_I and R_H after the final pull switch directions and respond as expected to shape anisotropy dominance, but R_I and R_H anisotropy is small. (c) The same effect is recorded for partially aligned γFe_2O_3. The angle associated with each point is a value of θ which indicates measurement direction with respect to alignment direction. These same R_I and R_H anisotropy trends are recorded for the shocked FeCu alloys (see Table 3). The particle size dependence for a given shock level indicated that shape deformation of the particles depends on particle size. In results presented in this paper, the maximum effect is noted for those alloys having precipitates in the 400–600 Å range.

$\gtrsim 800$ Å particles does such exist. In studies to date, the banding has only been clearly defined in the larger particles.

Consideration has been given to the variation of the hysteresis loop ratios with particle shape, particle size, and texture. LOTHIAN *et al.* (1958) have pulled wires containing the iron precipitates (see WASILEWSKI, 1974b, for summary). Hysteresis loop parameters were obtained for directions parallel and perpendicular with respect to the strain axis (Fig. 9a). In their work Lothian *et al.* noted an increase in transformation $\%$ with increase in wire elongation. In similar wire pulling experiments with a 9$\%$ Ni-Au alloy the initial response is magnetostrictive then shape effects are introduced (Fig. 9b), as would be expected, since the dominant anisotropy is due to $\lambda_s \sigma$. In this case the initial response is a decrease in R_I and an increase in R_{II}. The variation in R_I and R_{II} is also as expected for alignment of anisotropic particles of γFe_2O_3 (BATES, 1960-Fig. 9c).

Results presented in Table 3 for the shocked alloys are also in agreement with the work of LOTHIAN *et al.* (1958) (see Fig. 9a), i.e. R_I decreases and R_{II} increases in the perpendicular direction, the opposite being true for R_I and R_{II} in the parallel direction. The parallel and perpendicular designations being those with respect to an undirectional or uniaxial strain axis, alignment direction, or general anisotropy axis in a dispersed ferromagnetic system. The same generalized behavior can be also noted for natural materials in general where particle size and particle texture produced by oxidation are considered. However, in this latter case the angular variation should be isotropic in R_I and R_{II}. For a system which shows anisotropic R_I and R_{II} behavior it must also follow that the remanence is anisotropic. We have observed this in the shocked alloys, but more data are required to demonstrate the generalization.

7. Conclusions

Characteristics of first order shock transitions in iron and iron-nickel have been presented. If the iron is sufficiently large (say $\gtrsim 10~\mu$m) to resolve easily with an optical microscope, the shock induced microstructure is distinctive enough to be discriminated from microstructure associated with thermal mechanisms. Calibration results indicate that the mechanism of magnetization can be specified. Secondary effects associated with overprints by thermal transients can also be identified, but the effect itself is dependent on the Ni content of the alloy in question. If the temperature is high enough and (or) the time long enough, the primary shock structure can be altered significantly, but the recrystallization microstructure mimics any shock induced microstructure. For example, in a twinned grain the initial recrystallization will take place along twins and in the vicinity of inclusions if any are present.

Fine particles not observable by optical microscopy present a problem since characterization is difficult. Therefore the experiments with the FeCu alloys were designed to provide the experimental data base necessary to identify and discriminate dynamic and thermally induced magnetic effects in samples containing fine particle iron and iron-nickel.

The following conclusions are rendered as a result of our preliminary studies:

1) Thin foil electron microscopy confirms that the fcc precipitates exist with characteristic lobe contrast due to diffraction conditions in the strain field. Transformation induced by the shock impact causes the lobe contrast to disappear. This is due to the transformation. Dislocation tangles decorate the precipitates, and matrix dislocations often are observed to intersect the precipitates. Resolvable fine structure exists within the transformed precipitates.

2) Particle size, shape, and interparticle spacing can be controlled by varying the conditions of the precipitation anneal. The size and interparticle spacing can be arbitrarily varied by altering the composition (the iron) or the precipitation anneal.

3) The amount of transformed fcc iron, of a given size, increases as the pressure increases, and for a given peak pressure the amount transformed depends on particle size. The maximum shock induced anisotropy for a given pressure is observed in the (B3, A3) samples with a particle size between 400–600 Å. The anisotropy is based on hysteresis loop measurements parallel and perpendicular to the shock direction. Maximum magnetic hardness is also achieved in the same samples.

4) Initial estimates of the efficiency of remanence due to the shock transition have been made. Samples with a TRM will be partially demagnetized after a shock wave impact over the full particle size range studied ~ 250 Å–$\sim 1,000$ Å. The ratios of weak field remanence (RM), no matter what the origin, relative to the saturation remanence (SIRM) have been compared. Remanence associated with the shock tranformation always had a smaller ratio value. However uniaxial anisotropy is produced by the deformation, thereby imparting directional stability to the remanent state. The transverse components which are also observed are tentatively considered to exist because of the polycrystalline nature of the copper disc. Further directional control could be introduced by using single crystals or introducing a texture in the disc prior to the precipitation anneal.

5) Samples were shocked in three external field configurations: $H(+)$ applied in the direction of the shock normal, $H(-)$ applied in the opposite direction, and $H=0$ as measured with a fluxgate with $\pm 10\,\gamma$ resolution. In all cases where $H(+$ or $-)=0.5$ Oe was applied, the axial vector having the same sign as the field indicated that on the time scale of the transition, remanence was acquired in the same sense as the external field. When $H=0$, the sense of the axial component was that of the shock normal. *This apparent unidirectional aspect requires further experimental evaluation.*

6) An unusual characteristic in the (A3, B3) and (A2, B2) sample magnetic hysteresis is the recovery of saturation remanence after determination of the remanent coercive force. After H_R (the back field necessary to reduce remanence to zero) is determined, the field is reversed and from $H=0$ a field $H=H_R$ is applied and then reduced to zero at which point saturation remanence is recovered. This feature was observed in SB2A, SA21, SB3, SA31.

Most natural materials which can be classified as ferromagnetic dispersions contain a broad distribution of mixed anisotropies, and those artificial samples prepared from size graded Fe_3O_4 etc. can be expected to have clusters and inhomogeneous dispersion. In the FeCu samples the initial shape variation of the precipitate is zero. N is defined as $4\pi/3$, and the size range is narrow. K and λ_s are constants, and after transformation ΔN is uniquely a function of the average shape deformation of the spherical precipitates. Clustering is absent, and a mean particle spacing is specified by the nature of the iron content of the alloy and its thermal history. These features explain why magnetic hysteresis occurs over a limited field range and why the saturation remanence recovery as shown in Fig. 7 is possible. Quantification is in progress.

The saturation remanence recovery was not observed in fine grained ocean basalts, lodestone 99484 (see WASILEWSKI, 1977), or iron—Fe_3C containing Disko basalt.

7) Since the fcc→bcc transformation is a shock induced 1st order magnetic transition, an argument by analogy would suggest that the hcp→bcc′ transformation is similar, i.e. the remagnetization (bcc→hcp→bcc′ transformation) which takes place when $P > P_T$ (the pressure transition—see Fig. 2).

REFERENCES

ABUKU, S. and S. CHIKAZUMI, Magnetic study of martensitic transformation of stainless steel, *J. Phys. Soc. Japan*, **23**, 83–88, 1967.

BATES, G., Angular variation of the magnetic properties of partially aligned γ-Fe_2O_3 particles, *J. Appl. Phys.*, **32**, 2395–2405, 1960.

CAMPBELL, S.J. and P.E. CLARK, A mossbauer study of iron precipitates in CuFe alloys, *J. Phys. F: Met. Phys.*, **4**, 1073–1082, 1974.

CISOWSKI, S.M., J.R. DUNN, M. FULLER, M.F. ROSE, and P.J. WASILEWSKI, Impact processes and lunar magnetism, *Proc. Lunar Sci. Conf. 5th*, 2841–2858, 1974.

CISOWSKI, S.M., M.D. FULLER, Y.M. WU, M.F. ROSE, and P.J. WASILEWSKI, Magnetic effects of shock and their implications for magnetism of lunar samples, *Proc. Lunar Sci. Conf. 6th*, 3123–3141, 1975.

EASTERLING, K.E. and H.M. MIEKK-OJA, The martensitic transformation of iron precipitates in a copper matrix, *Acta. Metall.*, **15**, 1133–1141, 1967.

EASTERLING, K.E. and P.R. SWANN, The mechanism of nucleation of martensite in precipitates of iron in a copper matrix, *Proc. Conf. Mechanisms of Phase Transformations in Crystalline Solids, Inst. Metals*, 1968.

EASTERLING, K.E. and G.C. WEATHERLY, On the nucleation of martensite in iron precipitates, *Acta Metall.*, **17**, 845–852, 1969.

FOWLER, C.M., F.S. MINSHALL, and E.G. ZUKAS, *Response of Metals to High Velocity Deformation*, edited by P.G. Shewmon and V.F. Zackay, p. 275–308, Interscience, New York, 1961.

FULLER, M.D., Lunar magnetism, *Rev. Geophys. Space Phys.* **12**, 23–70, 1974.

FULLER, M.D., Magnetism of returned lunar samples, in *Proc. of Takesi Nagata Conference*, edited by R.M. Fisher, M. Fuller, V.A. Schmidt, and P.J. Wasilewski, NASA/GSFC, Greenbelt, Maryland, 1975.

HUGUENIN, R.L., The formation of goethite and hydrated clay minerals on mars, *J. Geophys. Res.*, **79**, 3895–3905, 1974.

JOHNSON, J.N. and R.W. ROHDE, Dynamic deformation twinning in shock loaded iron, *J. Appl. Phys.*, **42**, 4171–4182, 1971.

KUBO, H., Y. UCHIMOTO, and K. SHIMIZU, Martensitic transformation of iron precipitates in a copper—2 wt % iron alloy, *Met. Sci.*, **9**, 61–66, 1975.

LESLIE, W.C., D.W. STEVENS, and M. COHEN, Deformation and transformation structures in shock loaded iron-base alloys, in *High Strewgth Materials*, edited by V.F. Zackay, 382–435, John Wiley and Sons Ltd., New York.

LEWIS, M., Some experiments on synthetic titanomagnetites, *Geophys. J.R. Astron. Soc.*, **16**, 295–310, 1968.

LIVINGSTON, J.D. and J.J. BECKER, A study of precipitation hardening employing magnetic measurements, *Trans. Metall. Soc. AIME*, 316–319, 1958.

LOTHIAN, B.W., A.C. ROBINSON, and W. SUCKSMITH, Some magnetic properties of dilute ferromagnetic alloys II, *Philos, Mag.*, **3**, 999–1012, 1958.

MATSUURA, K., M. TSUKAMOTO, and K. WATANABE, The work hardening of Cu-Fe alloy single crystals containing iron precipitates, *Acta. Metall.*, **21**, 1033–1044, 1973.

NESS, N.F., K.W. BEHANNON, R.P. LEPPING, and Y.C. WHANG, *Interaction of Solar Wind with Mercury and Its Magnetic Field in Solar Wind Interaction with the Planets Mercury, Venus, and Mars*, edited by N.F. Ness, NASA SP-397, 1975.

NEWKIRK, J.B., Mechanism of precipitation in a Cu-2.5 Pct. Fe alloy, *J. Met.*, October 1957, 1214–1220, 1957.

PARRY, L.G., Magnetic properties of dispersed magnetite powders, *Philos. Mag.*, **11**, 303–311, 1965.

PHILLIPS, V.A. and S.D. LIVINGSTON, Direct observation of coherency strains in a copper-cobalt alloy, *Philos. Mag.*, **7**, 969–980, 1962.

SHARP, L.R., P.J. COLEMAN, JR., B.R. LICHTENSTEIN, C.T. RUSSELL, and G. SCHUBERT, Orbital mapping of the lunar magnetic field, *The Moon*, **7**, 322–341, 1973.

SMITH, C.S., Metallographic studies of metals after explosive shock, *Trans. AIME*, **214**, 574–589, 1958.

WASILEWSKI, P., Shock magnetization associated with meteorite impact at planetary surfaces, *The Moon*, **6**, 264–291, 1973.

WASILEWSKI, P., Magnetic remanence mechanisms in iron and iron-nickel alloys, Metallographic recognition centered and implications for lunar sample research, *The Moon*, **9**, 335–354, 1974a.

WASILEWSKI, P.J., Possible magnetic effects due to fine particle metal and intergrown phases in lunar samples, *The Moon*, **11**, 301–311, 1974b.

WASILEWSKI, P.J., Magnetism of meteorites, in *Proc. of Takesi Nagota Conf.*, edited by R.M. Fisher, M.D. Fuller, V.A. Schmidt and P. Wasilewski, NASA/GSFC, Greenbelt, MD, 1975.

WASILEWSKI, P.J., Shock loading meteoritic bcc metal above the pressure transition: Remanent-magnetization stability and microstructure, *Phys. Earth. Planet. Inter.*, **11**, 5–11, 1976.

WASILEWSKI, P.J., Magnetic and microstructural properties of the lodestone, Phys. Earth Planet. Inter., 1977 (in press).

WASILEWSKI, P.J. and A. DOAN, JR., Remanent magnetization and structural effects due to shock in natural and man-made iron-nickel alloys, in *Metallingical Effects at High Strain Rates*, edited by R.W. Rhode, B.M. Butcher, J.R. Holland, and C.H. Karnes Plenum Press, New York, 1973.

WASILEWSKI, P.J. and D. LARSON, Composition dependent magnetic and microstructural properties of bcc iron nickel spheres, *Phys. Earth Planet. Inter.*, 1977 (to be published).

WILLIAMSON, D.L., S. NASU, and V. GONSER, Surface states of Fe precipitates in Cu, *Acta. Metall.*, **24**, 1003–1008, 1976.

WINDOW, B., Precipitation in copper-iron alloys, *Philos. Mag.*, **25**, 681–699, 1972.

WOOLHOUSE, G.R., The mechanical properties of Cu—1 wt % Fe single crystals: An application of coherency strain control, *Philos. Mag.*, **30**, 65–83, 1974.

YIN, LO I., S. GHOSE, and I. ADLER, Investigation of a possible solar-wind darkening of the lunar surface by photoelectron spectroscopy, *J. Geophys. Res.*, **77**, 1360–1367, 1972.

YIN, LO I., T.S. TSANG, and I. ADLER, Electron spectroscopic studies related to solar wind darkening of the lunar surface, *Geophys. Res. Lett.*, **2**, 33–36, 1975.

Adv. Earth Planet. Sci., **1**, 147–168, 1977

The Thermoremanence Hypothesis and the Origin of Magnetization in Iron Meteorites

Aviva BRECHER and Lisa ALBRIGHT

*Department of Earth and Planetary Sciences, M.I.T.,
Cambridge, U.S.A.*

(Received June 20, 1977)

Iron meteorites are macroscopic single-crystals of Ni-Fe alloy (ave. 10% Ni), segregated into strongly ferro-magnetic (α) and weakly magnetic (γ) phases, intergrown in the octahedral system. Given their probable origin as molten metal cores or pods, slowly cooled (at rates ~ 1–$100°C/my$) in asteroidal bodies, they seem ideally suited to record ancient magnetic fields as thermal (TRM) or thermochemical (TCRM) remanent magnetization. To test this hypothesis, we investigated the intensity, relative stability and directional behavior in AF demagnetization of the natural (NRM), saturation (IRM_s), thermal (TRM) and spontaneous (SM) magnetization in several iron meteorites spanning the compositional-structural spectrum. The main results are: 1) The remanence intensity and relative stability increase systematically from the coarser to the finer-grained classes. The latter are capable of carrying a stable paleoremanence. 2) The NRM coercivity spectra, which are considerably harder than laboratory TRM's in the finest structured groups, gradually soften as grain-size coarsens. 3) All magnetization directions (NRM, TRM, SM) in octahedrites appear to be preferentially associated with the octahedral $\gamma\{111\}$ crystallographic planes on which $\alpha\{110\}$ plates nucleated and grew, and/or aligned with their intersections. The finer the structure, the clearer the link of magnetization directions to 'easy' crystallographic planes and axes. 4) A direct comparison of NRM and TRM demagnetization curves yields paleointensities in the range 0.3–3 Oe. However, the similarity of SM's (following zero-field cooling) to TRM's, implies fictitious ambient field values of 2–5 Oe. The extent to which SM mimics the TRM and NRM characteristics severely limits attempts to establish that the stable NRM in iron meteorites is an ancient TRM acquired in extraterrestrial fields of ~ 1 Oe. It is evident that the combined effect of magnetocrystalline and shape-anisotropy of the crystallographically ordered ferromagnetic kamacite (α)-phase had an overriding importance in producing the observed magnetic remanence characteristics of iron meteorites. Therefore, no reliable information regarding the presence, strength and sources of ancient solar-system fields can be retrieved from iron meteorites.

1.   Introduction

Laboratory studies of available extraterrestrial materials offer the opportunity both to test the ideas and advance the knowledge derived from years of terrestrial experience. In unraveling and interpreting the magnetic record of lunar rocks and meteorites, it is particularly important to become aware of any geocentric biases and to actually test assumptions which are often implicit in terrestrial rock magnetism (Fuller, 1974; Brecher, 1971, 1976a). The still unresolved issue of the origin of lunar magnetism has alerted many to the need for probing deeper and wider in the search for various magnetizing mechanisms operating on planetary surfaces, in planetary interiors and in space, rather than presuming a thermoremanence (TRM) (e.g., Wasilewski, 1973; Fuller, 1974; Dyal et al., 1974). The present work grew out of this new awareness: the iron meteorites as a group of well-characterized extraterrestrial objects apparently preserving a primary TRM (e.g., Guskova, 1972) seemed ideally suited to serve as a case study for verifying the validity of the TRM hypothesis, for identifying potentially important structural magnetic effects (Brecher, 1971) or unconventional mechanisms of magnetization (Brecher, 1976a). A brief background and justification follow:

The iron meteorites are typically gigantic single crystals of Ni-Fe alloy (5–20% Ni), which segregated on a macroscopic scale into ferromagnetic (α, BCC) Ni-poor kamacite and para- or (weakly ferro-) magnetic (γ, FCC) Ni-rich taenite. These phases are intergrown in a characteristically ordered octahedral structure: the $\alpha\{110\}$ planes nucleated epitactically on the $\gamma\{111\}$ faces and grew at the expense of the γ-host, by subsolidus diffusional equilibration during extremely slow cooling (~ 1–$100°C/my$) of initially molten cores or pods of Ni-Fe in differentiated asteroids. The basis for the prevalent structural classification and the accepted views on the thermal history and origin of iron meteorites can be found in several exhaustive references (e.g., Goldstein and Short, 1967; Scott and Wasson, 1975; Buchwald, 1975).

For the compositional range of interest (5–20% Ni) at least 80% of the octahedral structure formed in the temperature interval (780 to 620°C) just below the $\gamma \rightarrow \alpha$ transformation on the Ni-Fe phase equilibrium diagram (Goldstein and Ogilvie, 1965). This happens to coincide with the temperature range below the Curie point of the ferromagnetic α-phase (e.g., Hansen, 1958), in which a TRM could most effectively be blocked in the presence of a magnetic field, Since alloy phase separation and kamacite growth proceed simultaneously on cooling, a thermochemical remanence (TCRM) is a more correct assumption (Guskova, 1965 a, b). However, studies on growth-CRM in alloys and transformation-CRM in terrestrial rocks and oceanic basalts have indicated that a TCRM cooling has similar characteristics to that of TRM, if acquired during an initial cooling (Kobayashi, 1959; Marshall and Cox, 1971; Merrill, 1975). Thus, upon finding that a hard NRM component is generally present in iron meteorites (which is surprising in view of their coarsely crystalline structure), Soviet workers reasonably concluded that this NRM is probably a paleo-TRM or TCRM (Guskova, 1965 a, b; 1972, 1974). By comparing AF coercivity spectra and the thermal demagnetization of laboratory TRM's and of NRM, the Soviet scientists estimated that parent-body fields of 0.22–1 Oe (ave. 0.47 Oe) were needed during cooling below T_c to imprint the NRM as a TRM/TCRM (e.g., Guskova, 1972, App. 1). Based on similar magnetic evidence from other classes of meteorites (stony, stony-irons), Guskova and Pochtarev (1967) suggested that all meteorites originated in a core-mantle differentiated planetary body, capable of sustaining core-dynamo magnetic fields of 0.4–0.9 Oe.

In attempting to establish the nature and origin of magnetization in the iron meteorites each assumption must be explicitly stated and logically distinct aspects must be separated. In the present work we have first tried to examine if these objects are capable of preserving a stable NRM. Previous Russian studies on modelling of NRM (see GUSKOVA, 1972) suggested that some can. Our own study of magnetic domain structure in iron meteorites identified the microkamacite grains in the fine-grained plessite $(\alpha+\gamma)$ intergrowths as a very high-coercivity (SD) grain fraction, certainly capable of retaining a primary magnetic memory (BRECHER and CUTRERA, 1976). The plessite fields occur typically at the centers of the narrow residual γ bands, which decompose to an ordered $(\alpha+\gamma)$ intergrowth during the late stage of primary cooling $(T<500°C)$. They may also form upon mild reheating, by a two-stage diffusionless transformation $\gamma \rightarrow \alpha_2 \rightarrow \alpha+\gamma$ (BUCHWALD, 1975, vol. 1, p. 95 et seq.).

Further evidence of the capability to carry a stable remanence is adduced in Sec. 3.1 below. However, the validity of the common assumption that 'if they can, they do' is tested experimentally. The pitfalls are well-illustrated in Sec. 3.2. Another important question is whether any laboratory simulation can adequately model the NRM acquisition. The major drawback in trying to simulate in the laboratory the magnetization processes, is that the slow cooling over periods of maybe 0.5 to 50 million years, in cores of planetary bodies tens to hundreds of km in size, cannot be possibly duplicated. This intrinsic limitation may not be as severe as it seems; on the contrary it is best to minimally modify the structure of the meteorites under study (Sec. 3.3) while otherwise assessing the relative importance of structure for magnetics, as described below. Our primary objective was to determine if the NRM of iron meteorites is clearly a record of ancient magnetic fields, or is in any way an artifact of their characteristic crystallographic structure. Both Fe and Ni have magnetocrystalline anisotropy, their easy axes—[100] and [111], respectively—depending on the sign of the anisotropy constant K_1 (positive and negative, respectively) (CHIKAZUMI, 1964). For an alloy in the range of 0–20% Ni-Fe, K_1 is always positive, so kamacite should have three 'easy' [100] magnetization axes, along which the internal magnetization would stabilize in zero-field cooling (CHIKAZUMI, 1964). This was confirmed by FONTON (1960) who showed that in the iron meteorite Boguslavka (hexahedrite, 5.46% Ni)—a single crystal of kamacite—[100] is the easiest magnetic axis and [111] the hardest. Both the $[100]\alpha$ and the $[111]\gamma$ axes are contained in the $\alpha\{110\}$ set of planes, which coincide with the octahedral $\gamma\{111\}$ planes of the host taenite. These can be traced on an oriented meteorite section, as described in Sec. 2, and then related to the magnetization directions. Another possibly important factor is the shape anisotropy of the kamacite plates. The finer the meteorite structural class (based on kamacite bandwidth), the larger the expected contribution from shape anisotropy.

Table 1

Meteorite sample	Ni[1]	B.W.[1]	Cooling rate[3] °C/my	NRM (emu/ cm³)	MDF[4] (Oe)	% NRM at 100 Oe	IRM$_s$[5] (emu/ cm³)	IRM$_s$/ NRM	Meteorite control	NRM (emu/ cm³)
Babb's Mill (A) (Anom)	17.5	30 μm	—	0.0285	210	85	>3.81	>13.3	Babb's Mill	0.0720
Butler, Off, (An)	15.72	0.15	0.4	2.06	245	79	>5.41	>2.6	Butler	3.22
Smith's Mt. (Omf) (IIIB)	9.56	0.63	0.8–2	0.363	185	74	2.48	6.83	Smith's Mt.	0.0620
Carbo (Om) (IID)	10.15	0.85	1	0.327	18	14	2.58	7.89	Carbo	0.0234
Cosby's Creek (Og) (IA)	6.67	2.5	1–3	0.0274	15	12	0.48	17.5	Cosby's Creek	0.0265
Odessa (Ogg) (IA)	7.35	1.7	3	0.0434	15	6	0.747	5.4	—	—
Coahuila (H) (IIA)	5.59	H[2]	5[6]	0.144	45	19	0.93	6.4	Coahuila	0.138

 [1] Taken from Handbook of Iron Meteorites, 1975 (Buchwald); B.W.: Kamacite bandwidth.
 [2] Single kamacite crystal.
 [3] Determined by Goldstein and Short's (1967) rapid method.
 [4] Mean destructive field; alternating field value at which 1/2 NRM has been removed.
 [5] Saturation magnetization.
 [6] Randich, E. and J.I. Goldstein, *Meteoritics*, **10**, 479, 1975 (Abstract).

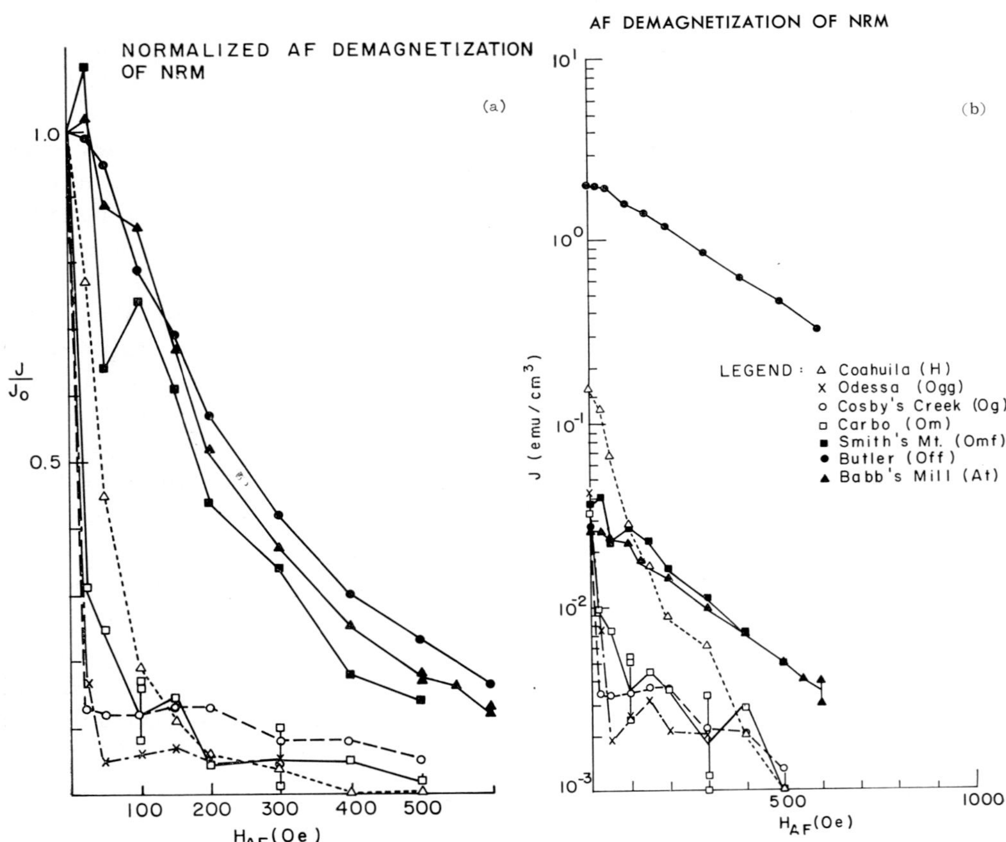

Fig. 1. (a) Fraction of NRM remaining as function of AF peak demagnetization field. (b) NRM, total moment (emu/cm³) as a function of demagnetizing field.

2. Experimental Methods

Samples of seven iron meteorites, representing a broad range of compositions and structures (Table 1) were obtained from the Harvard Museum (courtesy Prof. C. Frondel). Two adjacent cubes (~ 1 cm^3) of each, were cut with a bandsaw, preserving their mutual orientation. The magnetic moments were measured with a SSM-1A Schonstedt Spinner Magnetometer. Control samples were stored in zero-field and periodically remeasured. The other samples were progressively AF demagnetized in up to $\sim 1,000$ Oe peak field, using an instrument equipped with a 3-axis tumbler and with 3 sets of Helmholtz coils nulling the DC field at the sample. Absolute and relative intensities and directional changes of NRM were recorded (Figs. 1 and 2). The samples were then given a saturation remanence (IRM$_s$) in a 10 kOe field and

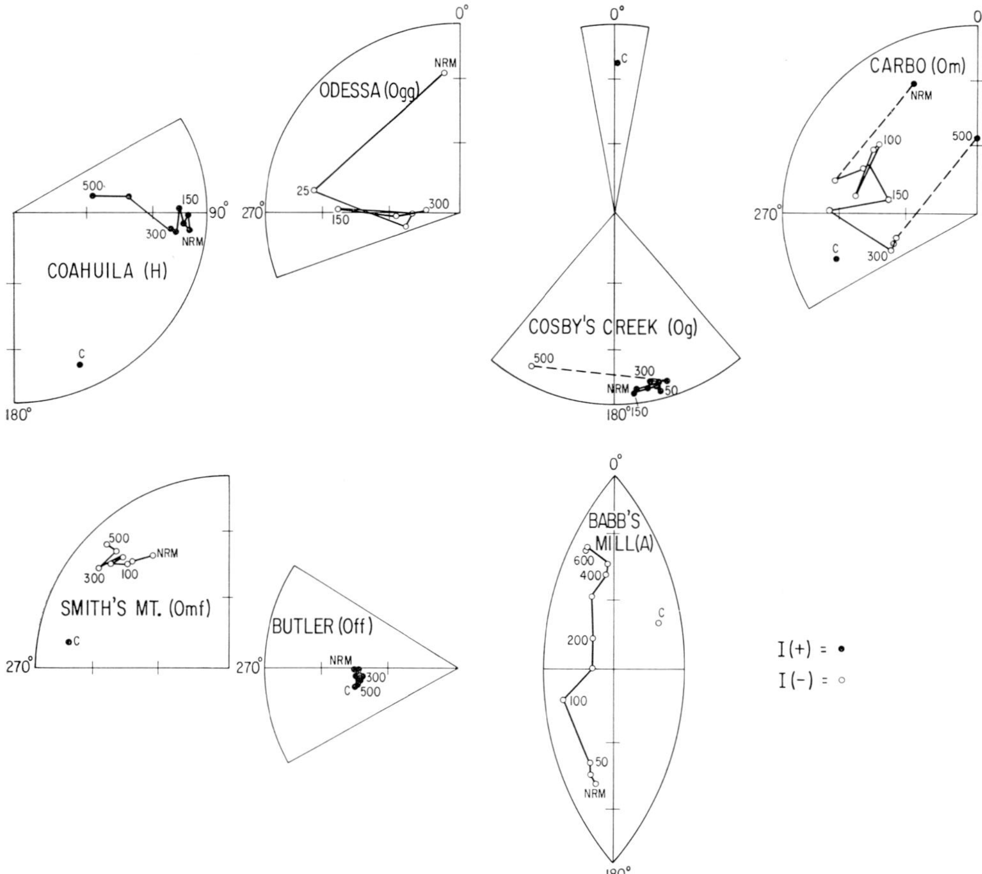

Fig. 2.   NRM directions of samples under AF demagnetization and of the controls (C). Positive inclination indicates lower-hemisphere vector. Three directions for Carbo at 100 Oe are two measurements 24 hours apart (directions close) and recleaning. Two directions at 300 Oe are two measurements of initial cleaning 24 hours apart.

the AF cleaning process repeated (Fig. 3).   Two orthogonal faces of each cubic specimen were ground, polished and etched to reveal the octahedral (Widmanstätten) pattern (Fig. 4).   To infer the crystal orientation of each specimen, we used the surface-trace-analysis method on macrophotographs of etched faces (Barrett and Massalski, 1966) (Fig. 4):   One cube face is used as the plane of projection (A) on

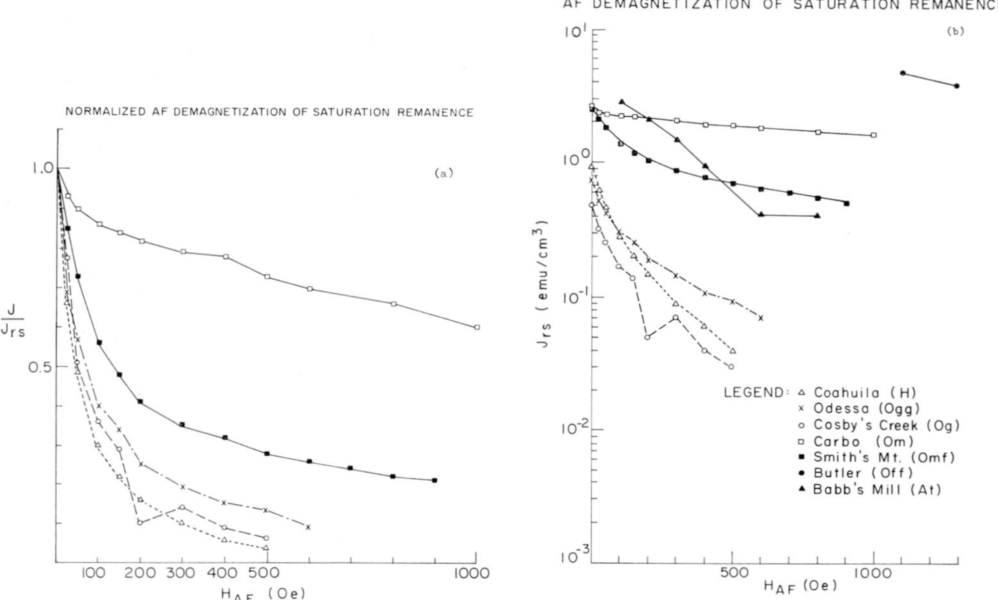

Fig. 3.   (a) IRM$_s$ microcoercivity spectra of five iron meteorites.   Absolute saturation remanence unknown for Babb's Mill and Butler (>5 emu which is limit of spinner). (b) Absolute scale IRM$_s$ values (emu/cm³) vs. applied peak alternating field.   A trend towards increasing single domain stability characteristics with decreasing kamacite bandwidth is indicated.

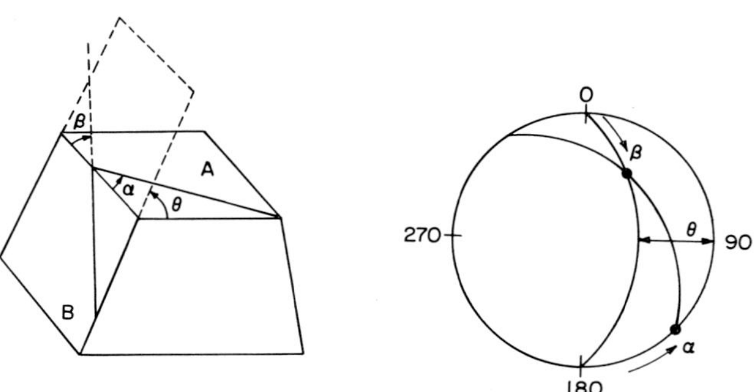

Fig. 4a.   Method of orienting kamacite planes in host taenite.   Plane A is the plane of projection on the stereonet.   Great circle connecting a tracing on each surface is the projection of the corresponding $\gamma\{111\}$ plane.

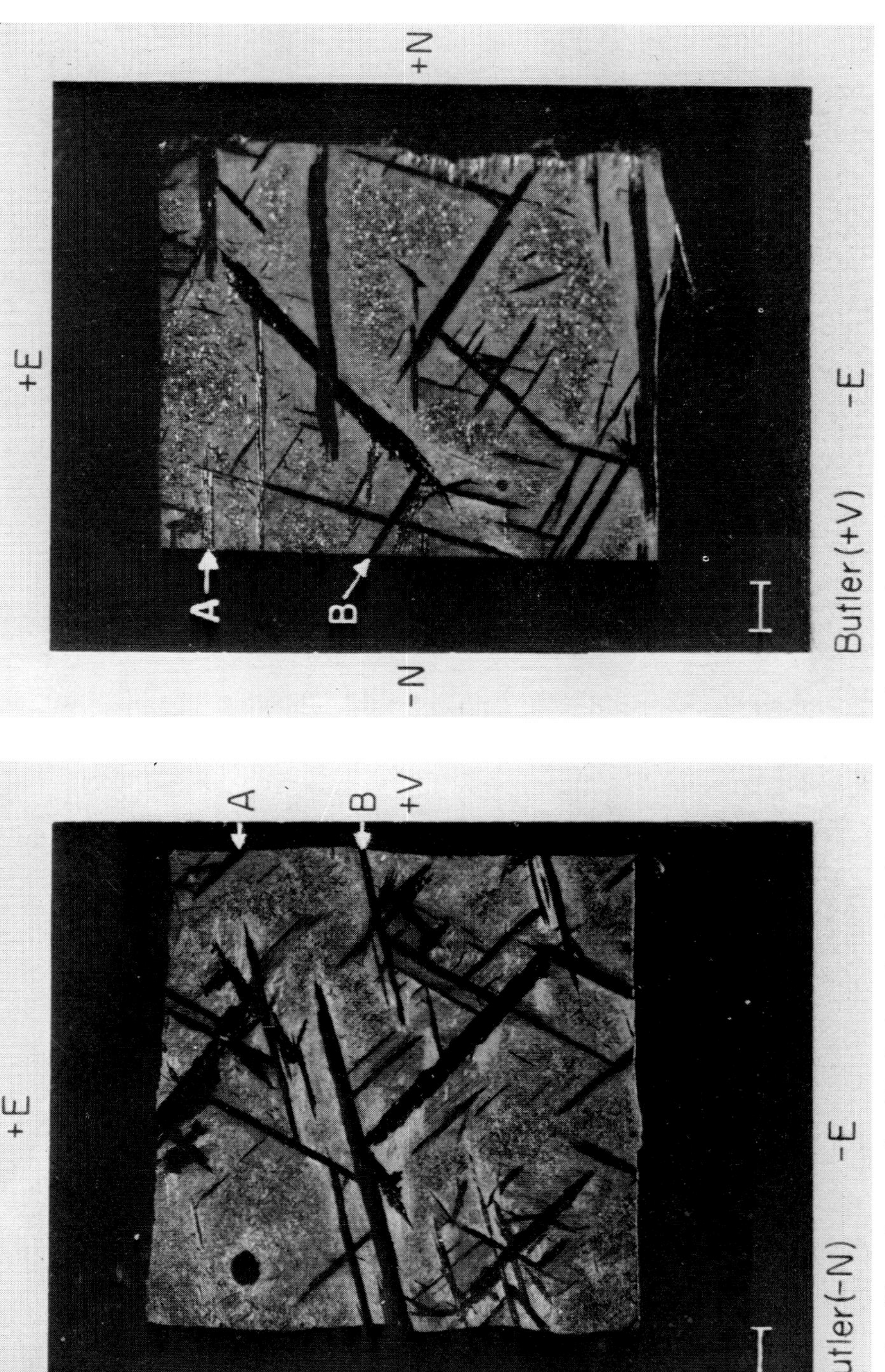

Fig. 4b.   Macrophotographs of Butler (sample).   A and B are tracings or kamacite planes which could be followed around the cube edge.

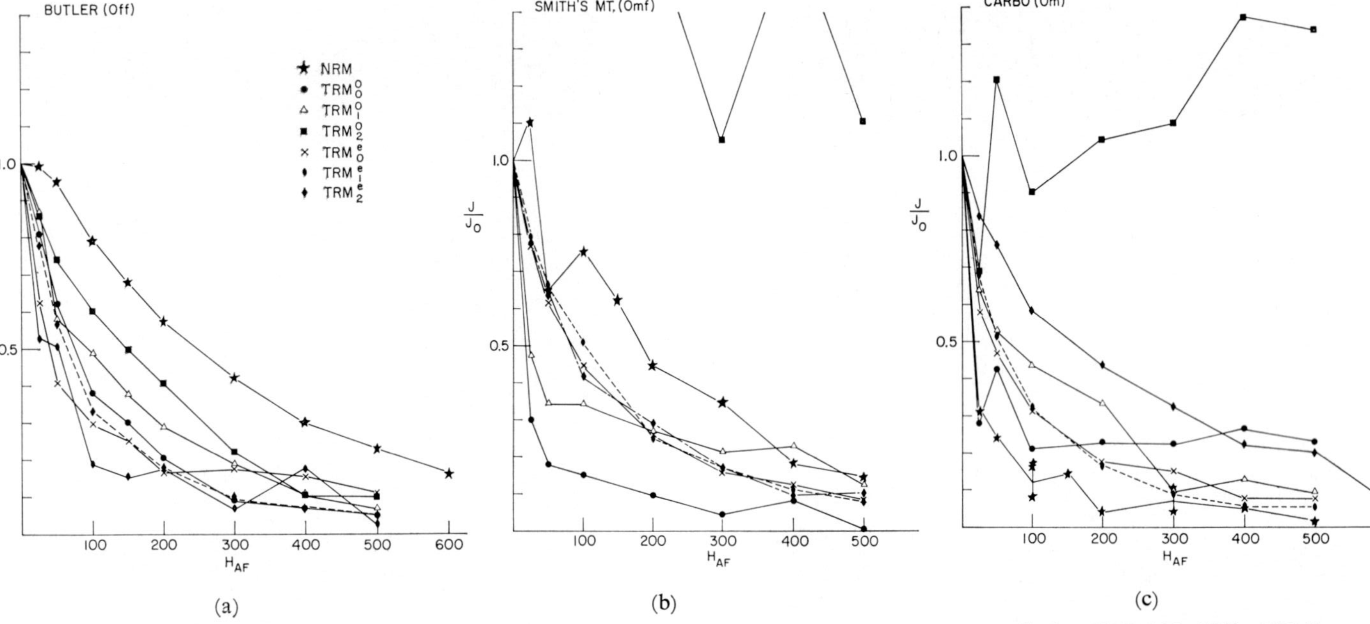

Fig. 5. Demagnetization characteristics of NRM, TRM^e and TRM^0 for (a) Butler, (b) Smith's Mt. (c) Carbo, (d) Babb's Mill. TRM's are listed in order of heating for (a)–(c). In (d), zero-field and earth's field coolings were staggered; cross-hatched area shows that the TRM^e's are remarkably similar so that no systematic changes in TRM^e coercivity spectra were induced by repeated heatings. TRM$_2^0$ of Babb's Mill, Smith's Mt. and Carbo show hard but unstable demagnetization curves, although error limits are acceptable. Within each group of TRM's (i.e., spontaneous and true) hardness increases with number of heatings.

NRM
TRMoo
TRMo1
TRMo2
TRMeo
TRMe1
TRMe2
2.15
1.97
BABB'S MILL (At)
TRMe
Jc/Jo
HAF(oe)
100   200   300   400   500

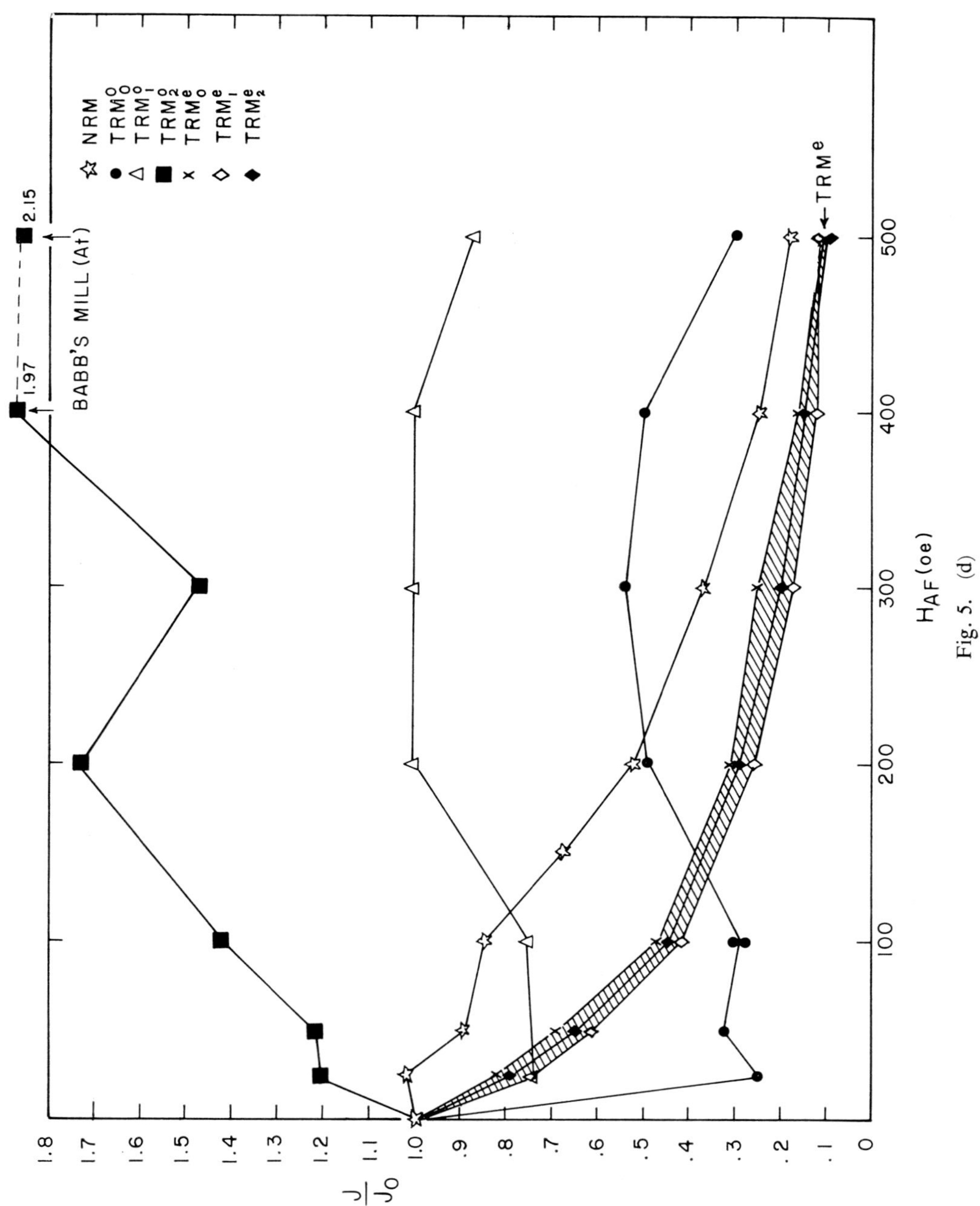

Fig. 5. (d)

a stereonet. The intersection of A with the orthogonal etched cube face B is the reference NS axis. The tracing of the kamacite plates on A with respect to this edge give declination markings on the perimeter of the stereonet. The angles which kamacite plates on face B make with this edge correspond to inclinations with respect to face A. The great circle connecting the traces of the same kamacite plate on both A and B is the projection of its plane on the stereonet (Fig. 4). This method was successfully applied to the fine to medium-grained (Off, Omf, Om) octahedrites with evident octahedral etch patterns (Wasson, 1974). Ambiguities in tracing $\gamma\{111\}$ planes were minimized by comparing the kamacite traces on the mutually oriented adjacent cubic samples of each meteorite. Traces identified on both polished cube

Table 2

Meteorite	$\mathrm{TRM}_0^0/\mathrm{NRM}$	$\mathrm{TRM}_1^0/\mathrm{NRM}$	$\mathrm{TRM}_2^0/\mathrm{NRM}$	$\mathrm{TRM}_0^e/\mathrm{NRM}$	$\mathrm{TRM}_1^e/\mathrm{NRM}$	$\mathrm{TRM}_2^e/\mathrm{NRM}$
Babb's Mill (A)	0.011	0.004	0.021	0.28	0.238	0.26
Butler (Off)	0.016	0.015	0.005	0.002	0.001	0.002
Smith's Mt. (Omf)	0.112	0.037	0.028	0.277	0.312	0.311
Carbo (Om)	0.068	0.125	0.007	0.198	0.210	0.096

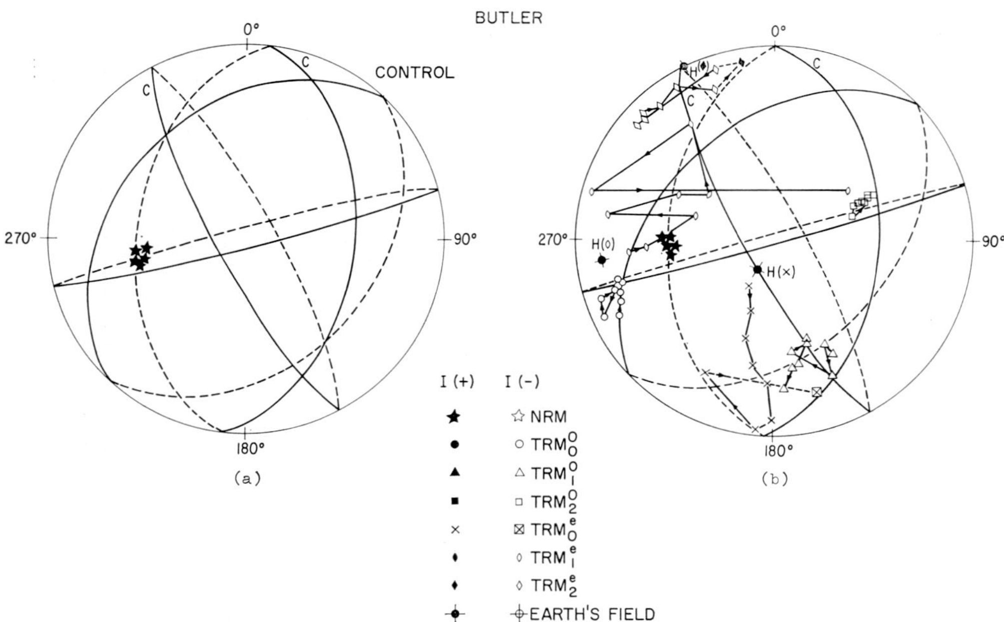

Fig. 6. (a) NRM and oriented kamacite planes in Butler (control). Solid lines are upper-hemisphere planes, dotted lines the lower-hemisphere complements. (b) NRM, TRM's and oriented kamacite planes in sample. Superscripts e, 0 indicate earth's- or zero-field for cooling; subscript indicates orientation of sample in space during cooling. "C" are high confidence planes mentioned in the text. NRM and all TRM's adhere to the kamacite planes.

faces are labeled 'C', for high-confidence planes in Figs. 6–8, whereas identifications based on different kamacite plates on each face yielded other possible sets of $\gamma\{111\}$ planes. By applying cubic symmetry as a constraint for angles between $\gamma\{111\}$ sets, we estimated our observational errors to always be $<20°$, whereas the errors in preserving the mutual alignment of the twin-cut cubic specimens were $\gtrsim 10°$. The structureless ataxite (A, Ni-rich) and hexahedrite (H, all kamacite) specimens, as well as the very coarse octahedrite (Ogg) Odessa specimen could not be absolutely oriented relative to principal cubic crystal axes. The polishing, etching and absolute orientation of specimens were performed after initial NRM, IRM_s and AF demagnetization measurements (Figs. 1–3), but prior to heating experiments. Then, they underwent a sequence of six brief (5 minutes) heatings at 800°C in a Schonstedt Thermal Demagnetizer (TSD-1). In the first three, they were cooled in zero field ($<10\gamma$) and in the last three, in known fields ($\gtrsim 1$ Oe), each in a different orientation with respect to the ambient field. To separate thermal effects from field effects, this sequence was altered only for the ataxite (Table 2), by alternating zero-field and field-coolings (Sec. 3.2). The sixth heating and cooling were carried out in the refractory chamber, to prolong the cooling intervals from 10–20 minutes to ~ 45 minutes. All these times are too short to allow for diffusional phase changes although

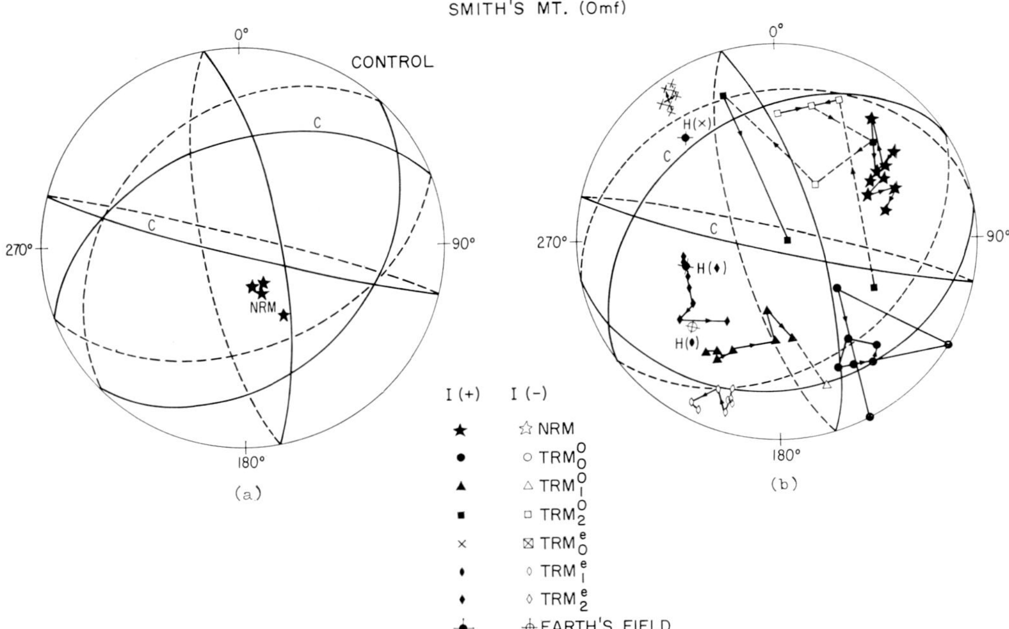

Fig. 7.   (a) NRM and oriented kamacite planes in Smith's Mt. (control). Plane of projection is different from that in Fig. 3 because previous plane was not a polished cube face. (b) NRM, TRM's and oriented kamacite planes in the sample. See Fig. 6 for explanation of scripts. All except NRM and TRM_2^e show some adherence to $\gamma\{111\}$, or equivalently, $\alpha\{110\}$ kamacite planes.

diffusionless massive transformation is possible (Sec. 3.3). Only negligible surficial oxidation occurred during heatings in air. After each cooling, the moments were measured and fully AF demagnetized (Figs. 5–8). Following thermal treatments, the samples were again saturated and AF cleaned to assess magnetic-structural modifications (Fig. 9). Finally, Van Zijl's method (COE and GROMMÉ, 1973) was used to estimate paleointensities, by comparing the coercivity spectra of NRM, TRM and SM (Fig. 10).

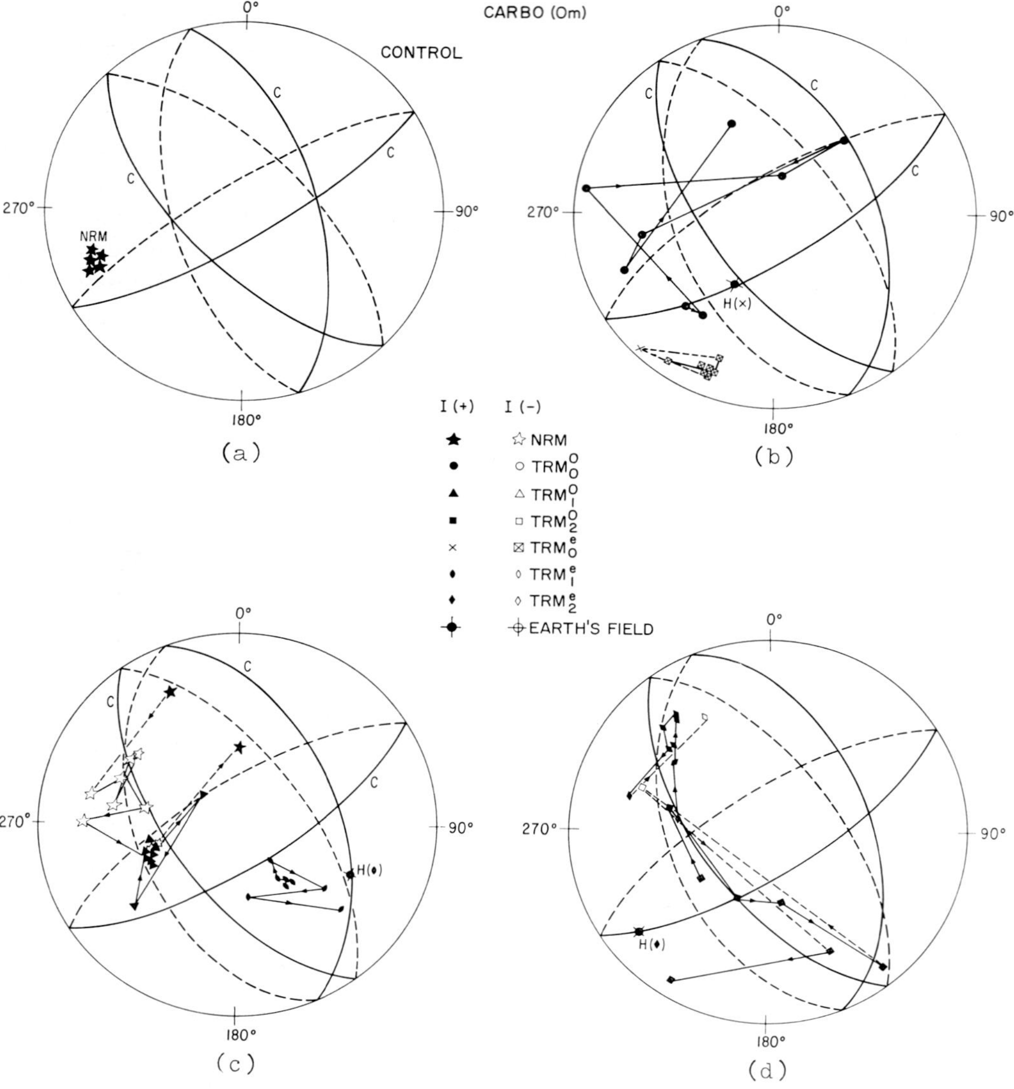

Fig. 8. (a) NRM and oriented kamacite planes in Carbo (control). (b), (c) and (d) NRM and TRM's plotted with projections of kamacite planes. NRM, TRM_0^0, TRM_0^1 and TRM_2^e show adherence to $\gamma\{111\}$ planes.

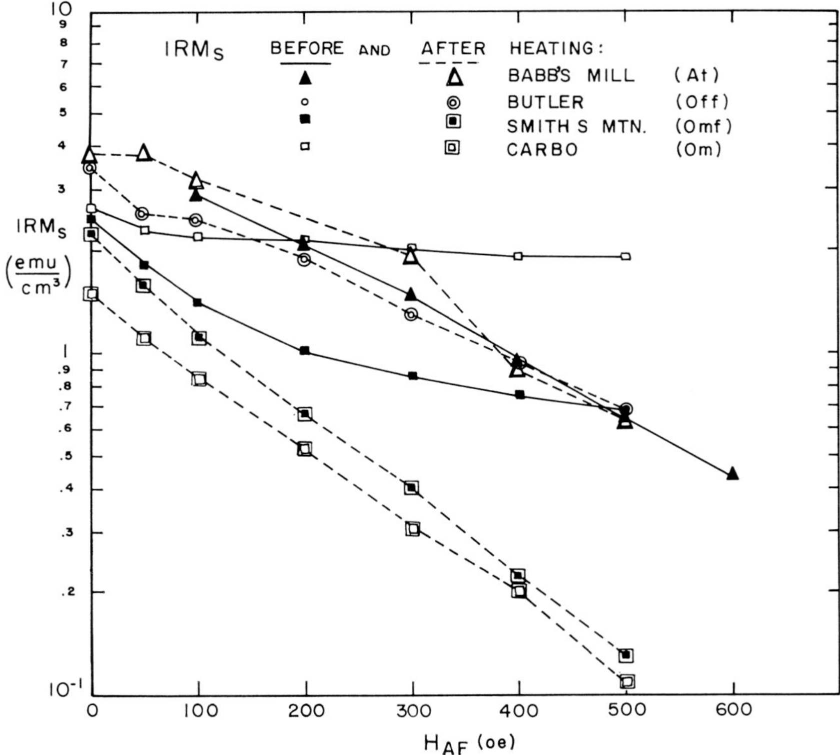

Fig. 9. Changes in IRM$_s$ intensity and microcoercivity spectra induced by the set of six heating cycles to 800°C; typically the IRM$_s$ value decreases and its spectrum softens as a result of heating. The curve for Butler before heating is not available (see Fig. 3 caption).

3.   Experimental Results and Analysis

3.1   NRM and IRM$_s$

The initial magnetic characteristics of the 7 iron meteorites under study are summarized in Table 1, together with their composition (Ni% contents), structural class cf. kamacite bandwidth (BW in mm) and metallographically inferred cooling rate (°C/my). Table 1 includes the NRM intensities for experimental and control samples. The NRM values are quite large and well clustered in the range $\sim 10^{-2}$–1 emu/cm³. Differences in NRM moments between subsamples of the same meteorite may be as large as a factor of ten and attest to nonuniform magnetization. The relative NRM stability to AF cleaning is expressed both as mean destructive field (MDF) and as fraction surviving ~ 100 Oe cleaning (T.1). To better characterize the magnetic structure, the IRM$_s$ moments and IRM$_s$/NRM ratios are also listed in Table 1.

During daily NRM measurements, over an 8-day period for the controls stored in zero-field, the NRM remained stable in both magnitude and direction (Fig. 2).

     A. Brecher and L. Albright

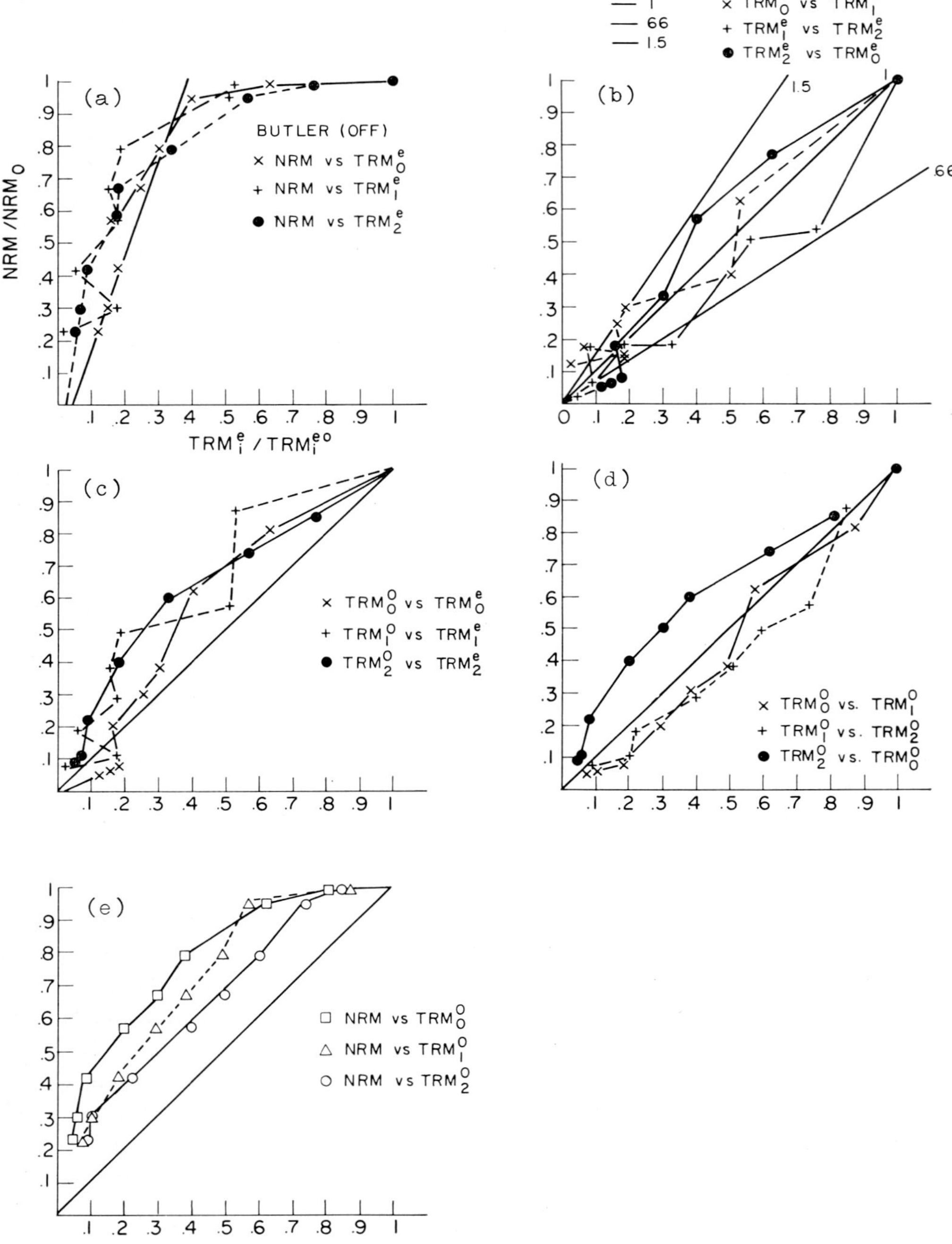

Fig. 10

For the samples group, the absolute NRM values and AF coercivity spectra are displayed in Fig. 1, with the corresponding directional changes shown in Fig. 2.

Two groups are resolved in Fig. 1, according to NRM stability: The harder group A is comprised of the ataxite (A), the finest (Off) and medium-fine (Omf) grained octahedrites. Group B shows a large soft NRM component and includes the coarser-grained meteorites: a medium (Om), coarse (Og) and coarsest (Ogg) octahedrite and the hexahedrite (H). Although the NRM stability clearly correlates with structural class, its intensity is a tradeoff between the grain-size of kamacite (coarsest in Off) and amount of kamacite (largest in H, which consists almost exclusively of α-phase, and smallest in A, which has little, though fine-grained, spindles of kamacite).

The same two magnetic groups are clearly resolved in the set of IRM_s demagnetization spectra (Fig. 2), though the Om meteorite is now seen to belong to the magnetically harder group A. The IRM_s spectra give clues to the capability for carrying a stable remanence, which again is clearly related to structural class, i.e., to the (effective) grain-size of the ferromagnetic kamacite. The IRM_s values are roughly an order of magnitude larger than the NRM, confirming the predominance of multi-domain (MD) carriers. The evident trend is that both the intensity and the relative stability of NRM and IRM_s increase from the coarser to the finer end of the kamacite-bandwidth spectrum.

With regard to the relative NRM directional coherence and stability, Fig. 1 shows it to also improve from the coarser-(H) to the finer (Off)-grained irons, excepting the ataxite (A). Several directional features are noteworthy: The NRM directions of sample and control coincide in the Off; have only similar inclinations in the H, Og, Om, and A; and are of opposite polarities in Omf. Sample and control NRM's lie in the same or in conjugate planes (small circles). Cleaned NRM directions may either cluster (as in Off, Omf and Og), or describe great-circle arcs (as for A and H), suggesting some type of planar confinement. The directional behavior does not correspond simply to the coercivity groups: The soft NRM components of Fig. 1 are apparently not secondary, as the directional clustering for H and Og indicate. Both soft and hard NRM components in Ogg and Om (Fig. 1) lack directional coherence (Fig. 2). The Off and Omf samples show high-coercivity and good directional convergence of NRM, so they may be assumed to carry a primary NRM. Further clues to the NRM directional behavior of octahedrites are obtained by reference to the principal crystallographic planes (Figs. 6–8): In the finest octahedrite (Off) Butler,

Fig. 10.   (a) These plots of NRM vs. TRM[e] residuals at each step of the AF cleaning process yield paleointensity estimates from the slopes (α) of the linear segments, assuming $\alpha = NRM/TRM = H_{anc}/H_{lab}$ (Table 3); (b) similar plots of paired TRM[e]'s provide an internal consistency check: expected slopes (0.66, 1 and 1.5) are indicated for comparison with three sets of data points; (c) by pairing zero-field and lab-field coolings, apparent field intensities of 2–5 Oe are obtained even in the absence of any external field (Table 3). The slope of 1 is given for reference; (d) the zero field moments are internally consistent and close to the reference slope of 1; (e) the NRM of Butler is considerably harder than any of the three zero-field moments; however, linear segments of $\alpha \sim 1$ may imply that the NRM resulted from primary cooling in zero-field!

the NRM directions of both control and sample coincide and line up with the lower hemisphere intersection of two $\gamma\{111\}$ octahedral planes (Fig. 6), to which they stay pinned during AF cleaning. In the medium-fine octahedrite (Omf) Smith's Mountain the NRM directions of both sample and control cluster well, near the midpoint of the three nearest lower hemisphere intersections of $\gamma\{111)$ planes (Fig. 7). In Carbo (Om), the control NRM lies on one $\gamma\{111\}$ plane, in the lower hemisphere; while the sample NRM falls on a different $\gamma\{111\}$ lower-hemispheric trace. Under AF cleaning, the sample NRM changes polarity twice, but always stays close to the same $\gamma\{111\}$ plane (Fig. 8). This behavior suggests that this $\gamma\{111\}$ plane is an easy plane of magnetization, but lacks an easy axis.

3.2 *Spontaneous (SM) and thermoremanent (TRM) magnetizations and their directional analysis*

Four meteorites, deemed capable of having preserved a stable paleoremanence, were subjected to thermal and magnetic cycling (Sec. 2). The first three heatings, followed by zero-field cooling, produced spontaneous moments (SM) labeled $\mathrm{TRM}^0_{0,1,2}$ in Table 2 and Figs. 5–8. The next sequence entailed cooling in laboratory fields of 0.5–0.85 Oe to imprint a thermoremanence $\mathrm{TRM}^e_{0,1,2}$. Only in the ataxite the sequence of zero- and field-coolings was staggered; results (Fig. 5d) showed that the order of the heating experiments is unimportant. However, the *presence* of an external field seems to influence the intensity of magnetization, although the moments do not vary in proportion to the applied field strength at cooling: The TRM^e values are typically 1/3–1/5 of the NRM intensity, whereas the SM values are one to two orders of magnitude smaller than the NRM (Table 2). However, note that for Butler (Off) the TRM^e and SM values are comparably low (1–0.1% of NRM). Moreover, for any two consecutive coolings in the same ambient field, the TRM may not change (e.g., in Omf and Om) or may decrease (e.g., in Off, Om and A). If the ambient field is varied from 0.5 to 0.85 Oe, the TRM value may stay the same (e.g., in Omf and Off) or even decrease (in A). Therefore, the field-proportionality assumption, crucial for the TRM hypothesis, is clearly not obeyed in these iron meteorites. The NRM is harder in AF demagnetization than all TRM^e's for the A, Off, and Omf irons, but considerably softer for Om. Within each group of TRM's (i.e., spontaneous and true), magnetic hardness increases with each consecutive heating cycle. The spontaneous magnetization coercivity spectra also become more jagged and apparently undemagnetizable (Fig. 5) (Brecher, 1976a). As will be shown below, this is only an *apparent* magnetic hardening, possibly due to planar pinning of SM dictated by internal magnetic anisotropy, since the IRM_s following the heating cycles are actually softer (Fig. 9).

The corresponding directional behavior of TRM^0 and TRM^e's (Figs. 6–8) is more revealing of the underlying magnetization mechanism than the AF coercivity spectra. Of the three octahedrites, whose magnetization directions are plotted relative to the observed traces of $\gamma\{111\}$ planes (Figs. 6–8), Butler (Off) shows most clearly significant results. Its three spontaneous moments ($\mathrm{TRM}^0_{0,1,2}$) cluster along the upper hemisphere intersections of pairs of $\gamma\{111\}$ planes. Their directional coherence is preserved in stepwise AF cleaning, just like that of NRM. By comparison, the $TRM^e_{0,1,2}$ show considerable directional scatter. The TRM^e polarity (in 2 out of 3 cases) differs from that of the ambient field at cooling, although initially they are roughly aligned with the field axis. With AF cleaning, the TRM^e's either oscillate near the $\gamma\{111\}$ plane closest to the direction of the ambient field or move about in a $\gamma\{111\}$ plane.

In the case of Smith's Mt. (Omf) (Fig. 7), all moments show some degree of association with a $\gamma\{111\}$ plane: they are directionally pinned to or migrating towards one. Now, the true TRM's show better directional coherence than the SM; upon repeated heatings the latter directions scatter increasingly and even change polarities and $\gamma\{111\}$ plane affiliation. Only one TRM^e direction coincides with that of the external field; the other two TRM^e cluster on the $\gamma\{111\}$ plane trace closest to the field direction, but with the proper polarity.

In Carbo (Om) all TRM's show greater directional scatter; only half ($TRM^{0;0;e}_{0;1;2}$) seem to be associated with a traced $\gamma\{111\}$ plane. TRM^e directions and polarities differ from those of the applied fields. However, just as for NRM, TRM reversals seem to take place within the same $\gamma\{111\}$ plane.

The overall trend evident from an analysis of directional change patterns is that the finer the structural class, i.e., the narrower the kamacite bands, the better defined is the planar confinement of magnetization to the $\gamma\{111\}$ planes, or to corresponding kamacite plates. This suggests that diminishing shape-anisotropy of kamacite plates relaxes the planar pinning of magnetization directions. Although not shown, all TRM's in the fine-grained ataxite are close to the same two conjugate great circles probably tracing $\gamma\{111\}$ planes, within which the NRM's of both sample and control lie in Fig. 2.

Both the intrinsic anisotropy and the external field direction influence the directionality of magnetization, the former clearly dominating in the fine-grained meteorites. In the Off, the smooth coercivity spectra and the better directional stability of SM's relative to true TRM's are due presumably to stronger magnetization pinning to the kamacite plates. According to standard stability criteria, the SM's produced in zero-field cooling of Butler (Off) mimic better the NRM than any of the TRM^e's. In the coarser-structured meteorites, increasingly harder and more jagged AF demagnetization profiles (Fig. 5) do not correspond to worse directional scatter, but are the effect of planar pinning and resistance to randomization of grain moments by alternating fields. This is supported by the data of Fig. 9, which indicate the net changes in the IRM_s coercivity spectra—and hence in effective magnetic grainsizes—produced by the six heating cycles.

3.3 Thermally induced changes

The changes in intensity of magnetization upon repeated heating (Table 2) might reflect loss of α-phase by $\alpha \to \gamma$ transformation on heating, followed by metastable survival of γ or by $\gamma \to \alpha_2$ transformation on cooling.

The IRM_s intensity has decreased somewhat in all heated Group A iron meteorites and their AF coercivity spectra are softer (Figs. 3 and 9). The finer-grained specimens (A, Off) have undergone only modest changes, whereas the coarser (Omf, Om) have lost a substantial fraction of high coercivity magnetic carriers ($H_{AF} > 100$ Oe). The lower the Ni-content and coarser the structure (Table 1), the larger are thermally-induced changes in the magnetic size-spectra. This trend could either be interpreted in terms of magneto-textural changes, such as annealing (e.g., of Neumann bands in kamacite) and grain growth (e.g., of microkamacite in plessite) upon repeated thermal cycling; or as loss of α-carriers related to the Ni-Fe phase transformations (GOLDSTEIN and OGILVIE, 1965; GOLDSTEIN and SHORT, 1967; BUCHWALD, 1975). Heating of kamacite to 800°C, into the γ-region, followed by rapid cooling probably leads to the formation of metastable martensite (α_2), whose magnetic properties are unknown. This supersaturated solution has the same chemical composition as the parent γ-phase (i.e., Ni-rich), so that it would have a lower magnetization than the original α-phase. Since the BCCα_2-plates have the same crystal habit as the BCC kamacite, the directional association of magnetization with these planes would probably not be affected. It is possible to explain the loss of the high-coercivity magnetic carriers, if we assume that the microkamacite in the Ni-rich plessite ($\alpha+\gamma$) regions transforms to γ-phase on heating, which survives metastably on cooling, thus no longer contributing to the IRM_s spectra. Indeed, BUCHWALD (1975, vol. 1, p. 94) shows that 10 minutes at 800°C suffice to homogenize submicroscopic ($\alpha+\gamma$) mixtures to taenite. All effects could account for the lower magnetization intensities upon heating (Table 2 and Fig. 9).

3.4 Paleointensity estimates: how meaningful?

In this paper, we have adopted the Van Zijl method for estimating paleointensities, in which the residual NRM fraction is compared to the respective TRM residual at each step of the AF cleaning process (COE and GROMMÉ, 1973). This method has been successfully applied earlier to derive paleofield strengths for other types of meteorites with an apparently stable and primary NRM component (BRECHER, 1972; BRECHER and RANGANAYAKI, 1975). The complete procedure is illustrated for Butler (Off) in Fig. 10. The paleofield values are obtained from the slopes (α) of the linear segments of normalized NRM vs. true TRM, assuming that the customary proportionality relation NRM/TRM$= H_{anc}/H_{lab}$ holds (Table 3). For the coarser Omf and Om irons, the average paleofields (1 and 0.34 Oe, respectively) are within the range

Table 3

Meteorite	$H_{ancient}$ (Oe)				H_{lab} (Oe) (using H_{anc} avg. value)		
	TRM_0^e	TRM_1^e	TRM_2^e	Avg.	TRM_0^0	TRM_1^0	TRM_2^0
Babb's Mill (A)	1.7	1.7	1.7	1.7	*	*	*
Butler (Off)	2	*	2.8	2.4	1.8	1.9	2.4
Smith's Mt. (Omf)	1.1	1.1	0.8	1.0	5	2	*
Carbo (Om)	0.34	0.34	0.52	0.43	*	*	*

*   Unable to obtain estimates for H, because of scatter and/or nonlinearity.

of GUSKOVA's (1965 a ,b) values. For the finer-grained A and Off, however, stronger fields (1.7 and 2.4 Oe, respectively) are obtained, in view of the higher AF stability of NRM relative to laboratory TRM's (Fig. 5). These estimates cannot be taken at face value because of the evidence discussed in Sec. 3.2, which suggests that the external magnetic fields do not simply determine either the intensity, direction, or the AF stability behavior of TRM, although the TRM^e intensity relative to NRM and TRM^0 may reflect the presence of a field (Table 2). Fictitious values for the ambient field are obtained even for zero-field cooling, if one solves for H_{lab} using $H_{anc}/H_{lab} = NRM/TRM^0$ (Table 3): these range from ~ 2 Oe to ~ 5 Oe for Butler (Off) (Fig. 10(e)). The internal consistency of these results, predicated on the existence of roughly linear segments in plots such as Fig. 10, is fair: Fig. 10(b) shows that the observed TRM^e slopes for paired TRM's follow roughly the expected slopes, within a tolerance factor of 1.5. Paired spontaneous moments TRM^0 also follow the expected slope (Fig. 10(d)). Finally, 10(c) and 10(e) confirm that apparent 'paleofields' for zero-field coolings exceed the values of both applied laboratory fields and of the estimated ancient fields (Table 3).

4.   Concluding Remarks

Magnetic fields are known to permeate space; be they of internal (planetary) or of external (solar) origin, they seem to have left some imprint on both terrestrial and extraterrestrial (lunar and meteoritic) rocks. Earlier work on deciphering the magnetic record locked in the most primitive meteorites available (the carbonaceous chondrites) indicated that relatively strong magnetic fields (~ 1 Oe), comparable to earth's field, were present when they formed as the solar system was coming into being (BRECHER, 1972; BUTLER, 1972; BANERJEE and HARGRAVES, 1972). In the present work we have carefully separated the two logically distinct questions of whether the iron meteorites are capable of preserving such a paleoremanence and whether they actually do. By standard reliability criteria used in terrestrial paleomagnetism, a 'prima facies' case could be made that iron meteorites carry an original paleoremanence of TRM/TCRM type. Upon careful probing, this hypothesis however plausible, is shown to be invalid. Our results show that, in fact, no reliable information on their early magnetic environments can be retrieved from iron meteorites. The natural (NRM), thermal (TRM) and spontaneous (SM) magnetization in iron meteorites are directionally influenced by the octahedral structure. The $\gamma\{111\}$ planes of the host taenite, on which the $\alpha\{110\}$ oriented plates of kamacite have nucleated and grown, are 'easy' preferred planes of magnetization, whose intersections define 'easy' magnetic axes. All the above types of magnetization tend to align with any one of these axes and to be confined ('pinned') to one or several of the preferred planes during progressive demagnetization. The laboratory thermoremanence (TRM) can align with an external magnetic field direction only if the field lies in or close to an octahedral plane, and diverges upon AF cleaning. The finer the crystalline structure, the stronger is its control of magnetization directions. Moreover, the relative stability of magnetization, whether acquired in cooling below the Curie point in zero-field (SM) or in earth's magnetic field (TRM), is comparable. Thus, spurious apparent paleofield intensities of ~1 Oe are obtained even for zero-field cooling. However, the magnitudes of true TRM are up to an order of magnitude higher than those of SM, excepting Butler (Off). TRM values are, in turn, considerably below the NRM levels. Therefore, even if the NRM intensity may be taken to indicate the presence of an external field during primary cooling, it is difficult to extricate any information on its strength and direction. Hence, the earlier conclusion of Soviet colleagues that iron meteorites, as pieces of planetary cores, carry the magnetic record of core-dynamo fields now appears doubtful. The strong and stable magnetization of iron meteorites may conceivably be merely spontaneous ferromagnetism, which has been directionally stabilized along and in energetically favorable 'easy' axes and planes defined crystallographically, through a combination of magnetocrystalline anisotropy and shape anisotropy of the kamacite needles and plates. These preferred directions may not necessarily be coherent on a larger scale (magnetic directions may differ in two adjacent cut cubes of an iron meteorite), so that asteroidal chunks of meteoritic iron may not be uniformly magnetized. However, the finer the metallographic structure, the stronger, stabler and more coherent the natural magnetization seems to become. The trend of increasing apparent paleointensity with decreasing kamacite bandwidth (Table 3) might be attributed to the higher spontaneous moments of finer grains and to more pronounced planar pinning.

There is some evidence that anisotropy effects associated with Ni-Fe carriers may be important in chondritic meteorites and in lunar rocks as well. Stacey *et al.* (1961) had noted a remarkable 'memory effect' for the stable, high-T (600–800°C) NRM component in chondrites, which reappeared, after repeated demagnetizations, with the same intensity and direction upon cooling from T_c. This is probably related to the pronounced anisotropy ($K_{\mathrm{max}}/K_{\mathrm{min}} \sim 1.4$–$2.1$) found in chondrites (ibid., Weaving, 1961). A similar memory effect and a strong magnetization following zero-field cooling from 800°C were reported in the lunar soil breccia 14301 by Dunn and Fuller (1972). Together with evidence for appreciable magnetic anisotropy and for texture-related directional confinement of NRM in a variety of lunar rocks (Brecher, 1976a, 1977), it seems probable that any extraterrestrial TRM may have been considerably modified or distorted by anisotropy effects.

Although it is disappointing that magnetic evidence from iron meteorites cannot be used with confidence to extract information on any ancient magnetic fields, it promises to provide a useful classification tool and may serve to confirm cogenetic groups of iron meteorites. Also, since these strong and ductile natural alloys are representative of an important class of man-made permanent-magnet materials, namely the diffusion-hardened alloys, it is possible to gain insight into the types and importance of magnetic anisotropy energies which control magnetic-field annealing of such alloys.

We thank Prof. C. Frondel for kindly providing us with samples from the Harvard Collection. This research was supported by a NASA grant (NSG 7199) to A.B. and by an M.I.T.-UROP award to L.A. Profs. R. Ogilvie and C. Smith generously shared with us their knowledge of the metallurgy of iron meteorites. Prof. J.I. Goldstein provided a helpful critique of an earlier draft of the manuscript. Mrs. D. Chouet is thanked for expert typing.

REFERENCES

BANERJEE, S.K. and R.B. HARGRAVES, Natural remanent magnetization of carbonaceous chondrites and the magnetic field in the early solar system, *Earth Planet. Sci. Lett.*, **17**, 110–119, 1972.

BARRETT, C.S. and T.B. MASSALSKI, *Structure of Metals*, 3rd ed., McGraw Hill, 1966.

BRECHER, A., On the primordial condensation and accretion environment and the remanent magnetization of meteorites, in *Evolutionary and Physical Properties of Meteoroids*, edited by C.L. Hemenway and A.F. Cook, NASA SP-319, 33, 311–329, 1971.

BRECHER, A., Memory of early magnetic fields in carbonaceous chondrites, in '*On the Origin of the Solar System*', *Proc. CNRS Symposium*, 260–273, 1972.

BRECHER, A., Textural remanence: a new model of lunar rock magnetism, *Earth Planet. Sci. Lett.*, **29**, 131–145, 1976a.

BRECHER, A., Meteoritic magnetism: implications for parent bodies of origin (invited review for Proc. IAU. Coll. #39, Lyon, France, Aug. 1976), in *The Interrelated Origins of Meteorites, Asteroids, and Comets*, edited by A.H. Selsemme, 1976b (in press).

BRECHER, A., Interrelationships between magnetization directions, magnetic fabric and oriented petrographic features in lunar rocks, *Proc. Lnnar Sci. Conf. 8'th*, 1977 (in press).

BRECHER, A. and M. CUTRERA, An SEM study of the magnetic domain structure of iron meteorites and their synthetic analogues, *J. Geomag. Geoelectr.*, **28**, 31, 1976.

BRECHER, A. and R.P. RANGANAYAKI, Paleomagnetic systematics of ordinary chondrites, *Earth Planet. Sci. Lett.*, **25**, 57, 1975.

BUCHWALD, V.F., *Handbook of Iron Meteorites*, Univ. of Calif. Press, 1975.

BUTLER, R.F., Natural remanent magnetization and thermomagnetic properties of the Allende Meteorite, *Earth Planet. Sci. Lett.*, **17**, 120, 1972.

CHIKAZUMI, S., *Physics of Magnetism*, John Wiley and Sons, Inc., 1964.

COE, R.S. and C.S. GROMMÉ, A comparison of 3 methods of determining geomagnetic paleointensities, *J. Geomag. Geoelectr.*, **25**, 415, 1973.

DUNN, J.R. and M. FULLER, On the remanent magnetism of lunar samples, with special reference to 10048, 55 and 14053, 48, *Proc. Lunar Sci. Conf. 3rd*, 2363, 1962.

DYAL, P., C.W. PARKIN, and W.D. DAILY, Magnetism and the interior of the Moon, *Rev. Geophys. Space Phys.*, **12**, 568, 1974.

FONTON, S.S., Measurements of reversible magnetic susceptibility in the principal crystallographic directions of Ni-Fe crystal, *Kristallografiya*, **5**, 325, 1960 (in Russian).

FULLER, M., Lunar magnetism, *Rev. Geophys. Space Phys.*, **12**, 23, 1974.

GOLDSTEIN, J.I. and R.E. OGILVIE, The growth of the Widmanstätten pattern in metallic meteorites, *Geochim. Cosmochim. Acta*, **29**, 893, 1965.

GOLDSTEIN, J.I. and J.M. SHORT, The Fe-meteorites, their thermal history and parent bodies, *Geochim. Cosmochim. Acta*, **31**, 1001, 1967.

GUSKOVA, E.G., Study of natural remanent magnetization of iron and stony-iron meteorites, *Geomagn. Aeron.* (English transl.), **V**, #1, 91–96, 1965a.

GUSKOVA, E.G., The nature of the natural remanent magnetization of meteorites, *Meteoritika*, **XXVI**, 60, 1965b (in Russian).

GUSKOVA, E.G., Magnetic properties of meteorites in the collection of the Committee on Meteorites of the USSR Academy of Sciences, *Meteoritika*, **XXX**, 74, 1974 (in Russian).

GUSKOVA, E.G. and V.I. POCHTAREV, Magnetic fields in space according to a study of the magnetic properties of meteorites, *Geomagn. Aeron.*, **7**, 245–250, 1967.

GUSKOVA, YE. G., *The Magnetic Properties of Meteorites* (English transl.), Nauka, Leningrad, NASA-TT-F-792, 1972.

HANSEN, M., *Constitution of Binary Alloys*, Handbook, McGraw Hill, Inc., 1958 and R. Elliott's supplement, McGraw Hill, Inc., 1965.

KOBAYASHI, K., Chemical remanent magnetization of ferromagnetic minerals and its application to rock magnetism, *Geomag. Geoelectr.*, **10**, 99, 1959.

MARSHALL, M. and A. COX, Effect of oxidation on the natural remanent magnetization of titanomagnetite in sub-oceanic basalt, *Nature*, **230**, 28, 1971.

MERRILL, R.T., Magnetic effects associated with chemical changes in igneous rocks, *Geophys. Surv.*, **2**, 277, 1975.

SCOTT, E.R.D. and J.T. WASSON, Classification and properties of iron meteorites, *Rev. Geophys. Space Phys.*, **13**, 527, 1975.

STACEY, F.D., Paleomagnetism of meteorites, in *Ann. Rev. Earth Planet. Sci.*, vol. 4, p. 147, 1976.

STACEY, F.D., J.F. LOVERING, and L.G. PARRY, Thermomagnetic properties, natural magnetic moments and magnetic anisotropies of some chondritic meteorites, *J. Geophys. Res.*, **66**, 1523, 1961.

WASILEWSKI, P.J., Shock remagnetization associated with meteorite impact at planetary surfaces, *The Moon*, **6**, 264, 1973.

WASSON, J.T., *Meteorites: Classification and Properties*, Springer-Verlag, New York, 1974.

WEAVING, B., Magnetic anisotropy in chondritic meteorites, *Geochim. Cosmochim. Acta*, **26**, 451, 1961.

Adv. Earth Planet. Sci., **1**, 169–178, 1977

Thermal Overprinting of Natural Remanent Magnetization and K/Ar Ages in Metamorphic Rocks

Kenneth L. Buchan,* Glenn W. Berger,* Michael O. McWilliams,**
Derek York,* and David J. Dunlop*

*Geophysics Laboratory, Department of Physics, University of Toronto, Toronto, Canada
**Research School of Earth Sciences, Australian National University, Canberra, Australia

(Received July 2, 1977)

Burial metamorphism generates secondary components of NRM (natural remanent magnetization). Often two or more remanences are found superimposed in a single geological unit. We trace the steps necessary to resolve and date such magnetic overprints in three studies of intensely metamorphosed rocks from the Grenville Province of the Canadian Precambrian Shield. First, it is shown that the various components may be associated with different rock-forming minerals and can be separated by orienting aggregates of dark or light grains in coarse-grained samples. Next, these magnetizations are shown to have originated at least in part as TRM's (thermoremanent magnetizations). Their relative ages can then be assigned on the basis of blocking temperatures. Finally, we estimate their absolute ages using the $^{40}Ar/^{39}Ar$ stepwise heating technique, taking into account the effects of a long heating-cooling history on both the magnetic and radiometric systems.

1. Introduction

Recent paleomagnetic studies of North American Precambrian rocks (e.g., Buchan and Dunlop, 1973, 1976; Irving *et al.*, 1974; Fahrig *et al.*, 1974; Ueno *et al.*, 1975; Roy and Lapointe, 1976; Irving and McGlynn, 1976; Schwarz, 1977; Vitorello and Van der Voo, 1977; Buchan, 1977a) have revealed numerous examples of superimposed natural remanent magnetizations (NRM's). One of the component NRM's may be primary, but the other component(s) are thermal or chemical overprints resulting from burial metamorphism. Frequently, all the components are overprints.

Paleomagnetic techniques for separating multiply overprinted NRM's are just being developed. Determining reliable paleomagnetic directions for the components—a difficult task if their coercivity and blocking temperature spectra overlap—is not enough. Determining the origins of the components, whether TRM or CRM, is an obligatory first step in assigning relative, and ultimately absolute, ages to the components. Age determination is almost taken for granted in single-component studies where a primary NRM can be assumed, but it is far from trivial in rocks that have been repeatedly remagnetized. Indeed it is the principal difficulty in establishing apparent polar wander paths (APWP's) for Precambrian time.

This paper describes some novel approaches to the problems of metamorphic

(a)
T(°C)
J/J_N
BARK LAKE
WHOLE-ROCK
420
thermal
af
H(Oe)
N
660
638
150 200 300 400 800
100
50 250 600
20 1000
200 600
350 525

(b)
J/J_N
BARK LAKE
MINERAL SEPARATE
420-A
H(Oe)
N
25
100
50 200-1000
150

(c)
J/J_N
BARK LAKE
MINERAL SEPARATE
420-C
H(Oe)
N
300
800 200 600
250
400
0-50 150
100
75
1000

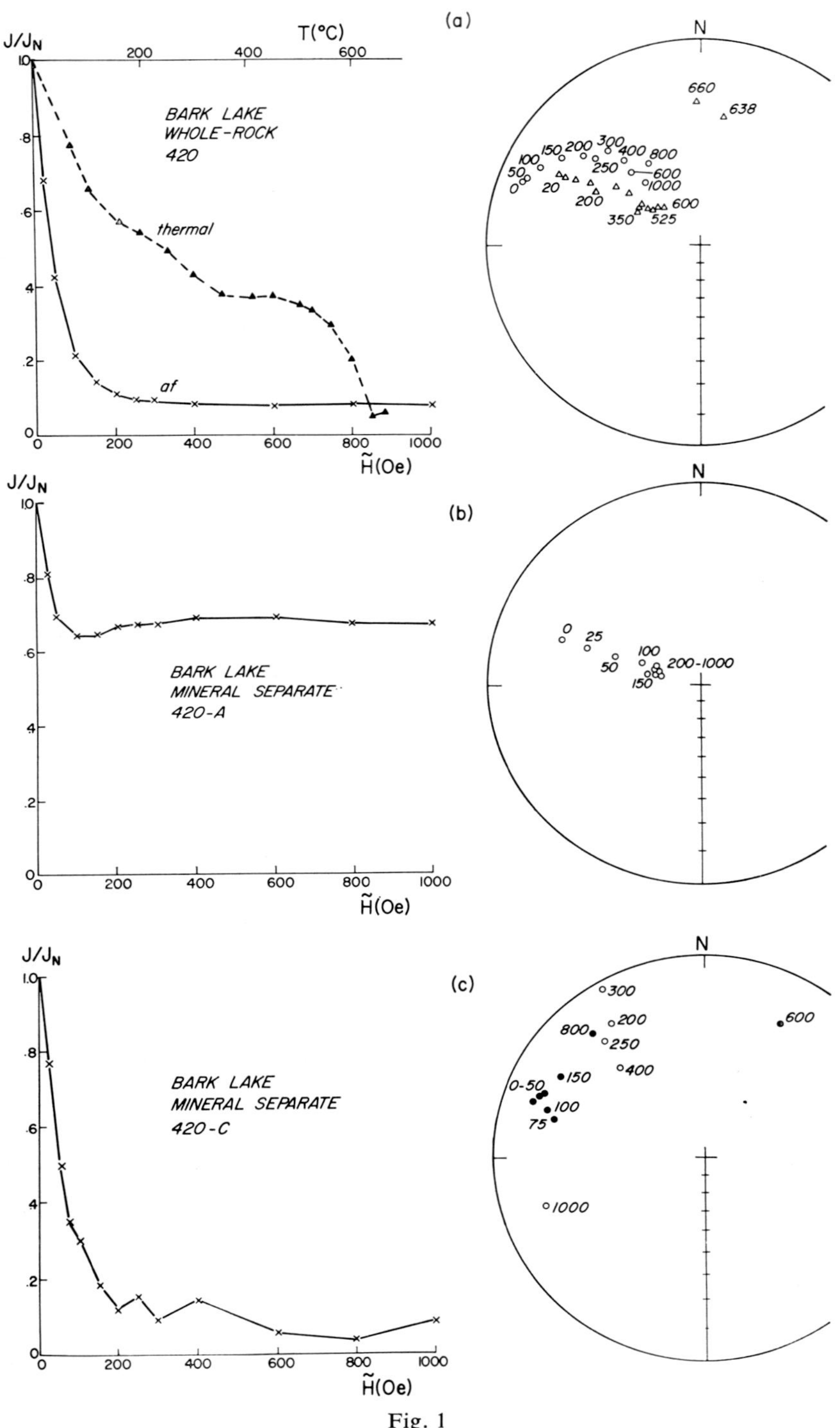

Fig. 1

overprinting. Our examples are drawn from the Grenville structural province of the Canadian Precambrian Shield, where regional metamorphism was unusually severe, and uplift and cooling were unusually slow, spanning 100 to 200 my. Grenville rocks frequently exhibit superimposed NRM's in individual samples.

As our case histories will make clear, a complete multi-component paleomagnetic study demands time and ingenuity. On the other hand, a successful study of this kind determines an APWP in miniature for the body studied, rather than a single paleopole. In what follows, magnetic studies are due to K.L.B., M.O.McW and D.J.D. and $^{40}Ar/^{39}Ar$ work is by G.W.B. and D.Y.

2.   Resolving Superimposed Magnetizations by Bulk Mineral Separation

The isolation of magnetic components superimposed in individual specimens has usually been attempted using the indirect approaches of vector subtraction (ZIJDERVELD, 1967; BUCHAN and DUNLOP, 1976) and the intersection of remagnetization circles (HALLS, 1976). We are currently attempting a direct separation of components in samples of the coarse-grained Bark Lake diorite (one of the Haliburton intrusions of BUCHAN and DUNLOP, 1976), employing magnetic microanalysis (WU *et al.*, 1974).

Consider, for example, the sample of Fig. 1. In whole-rock analysis (Fig. 1a), the direction of magnetization moves progressively on both alternating field (AF) and thermal demagnetization from a shallowly inclined Grenville *B* direction toward a steep *A* direction. (The *A* and *B* directions for the Haliburton intrusions were established by BUCHAN and DUNLOP (1976) utilizing stable endpoints and vector subtraction respectively.) In Fig. 1a, directional coherence diminishes at the higher alternating fields, presumably because the *A* component constitutes a small fraction (7%) of the NRM intensity (1.8×10^{-3} emu/cm^3).

The sample of Fig. 1 is composed of rather large aggregates of dark rock-forming minerals (largely hornblende and biotite, the latter often altered to chlorite) set in a groundmass of both light (feldspar) and dark (biotite, hornblende and pyroxene) minerals. We oriented blocks of these bulk mineral fractions whose magnetizations are sufficiently intense to be measured with a standard spinner magnetometer. These separates are 0.1 to 2 cm^3 in volume, larger than the oriented grains cut from thin sections by WU *et al.* (1974).

We find that the *A* and *B* magnetizations are preferentially associated with the different mineral fractions of the whole rock. The *B* component (Fig. 1c) is concentrated in the large aggregates of dark minerals where titanomagnetites are observed

Fig. 1.   AF and thermal demagnetization of a sample of the Bark Lake diorite (one of the Haliburton intrusions) showing superimposed *A* and *B* magnetizations. Directions are plotted on a stereographic projection. Closed and open symbols indicate positive and negative inclinations respectively. Circles and crosses denote AF results, triangles denote thermal results. Numbers indicate peak AF's in Oe or temperatures in °C. Characteristics of (a) a whole-rock specimen, and oriented bulk separates from (b) groundmass material and (c) a dark mineral segregate are compared.

to be the dominant opaque mineral. The *A* component, on the other hand, resides preferentially in the groundmass fraction where both magnetite and hematite have been identified in polished thin section.

The *A* component intensity (Fig. 1b) constitutes nearly 70% of the NRM (0.66×10^{-3} emu/cm³) of the groundmass separate. The *A* endpoint direction is now attained 100 Oe, whereas the magnetization vector of the whole-rock sample had barely begun to swing in 100 Oe.

No evidence of an *A* component is observed in the dark mineral segregate (Fig. 1c). The magnetization direction remains constant to 150 Oe, then moves to somewhat higher declinations, before becoming random at 600 Oe and above. The directional shift between 150 and 400 Oe may be due, in part, to problems encountered in measuring the low-intensity magnetization of the small separate.

Nevertheless, this example demonstrates the usefulness of direct mineral separation to isolate the directions of superimposed magnetizations in coarse-grained rocks. This method is applicable to several problems in the study of multi-component magnetizations:

a)   confirming whole-rock endpoints which may have been established by indirect methods such as vector subtraction and intersection of remagnetization circles;

b)   isolating endpoints when indirect methods fail; and

c)   determining the magnetic mineral carriers and their association with rock-forming minerals.

This latter application is potentially important in establishing the origin and relative ages of superimposed components.

3.   Problems in Establishing Relative Ages of Superimposed Magnetizations

Isolating endpoint directions constitutes but the first step in the study of multi-component remanences. Next, one must date each component. We will consider this problem in two steps: first, establishing the *relative* ages of the various components by determining their origins, and second, establishing their *absolute* ages by radiometric dating.

3.1   Relative ages of thermal and chemical overprints

There is a widely-held misconception that relative ages of superimposed magnetizations can be defined more or less automatically by examining their blocking temperature (T_B) spectra. The component with the highest blocking temperatures is deemed to be the oldest, while successively lower T_B components are assumed to be progressively younger.

This simple picture presupposes that all components are partial TRM's, primary or secondary. A minimum test of this assumption is that, in comparing samples from the same site or neighbouring sites, the blocking temperature ranges of the various components should not overlap. Even with this proviso, it is easy to imagine anomalies, e.g., a CRM overprint carried by hematite with apparent blocking temperatures

(apparent, because the overprint is chemical, not thermal) between, say, 600 and
670°C that postdates a magnetite TRM with $T_B \leq 580$°C. Blocking temperature data
alone are not sufficient to prove the thermal origin of overprints.

3.2 Nature and relative ages of Grenville components

The necessity for detailed magnetic studies to establish relative ages is illustrated
by the Haliburton (BUCHAN and DUNLOP, 1976) and Thanet (BUCHAN, 1977a) intrusives. In each study, a steeply inclined Grenville A direction and a shallow B direction have been isolated (Fig. 2). Note that the B directions are of the same polarity
in the two intrusives but the A directions are reversals of each other, being steeply
upwards in the Haliburton and steeply downwards in the Thanet.

The Haliburton magnetization vector swings from the B towards the A direction
on both AF and thermal demagnetization (Fig. 2a). Other published Grenville studies
(IRVING et al., 1974; UENO et al., 1975) show the same pattern, which has been taken
to be characteristic. On a simple thermal overprinting model, the A magnetization,
since it has higher T_B's, is older than the low T_B B component.

However, results from the Thanet gabbro do not agree (Fig. 2b). Upon AF demagnetization, the directional swing is from B towards A, but on thermal demagnetization, the vectors always move away from the A direction towards the B direction.
That is, the B component, although it is 'soft' relative to the A component upon AF
demagnetization, has higher blocking temperatures than the A component.

If we assume that the Haliburton A and Thanet A components are essentially
contemporaneous, as are the Haliburton B and Thanet B magnetizations, the assumption of thermal overprinting cannot be correct in both formations. Additional evidence is necessary to assign relative ages to Grenville A and B components.

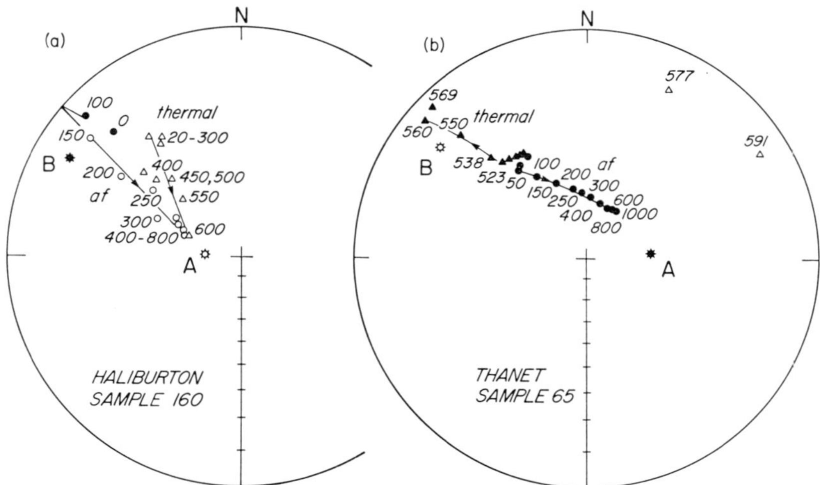

Fig. 2. Comparison of AF and thermal demagnetization characteristics of samples showing superimposed ($A+B$) magnetizations from (a) the Haliburton intrusions and (b) the
Thanet gabbro. Symbols as in Fig. 1. Best-fitting great circle paths are superimposed on
the data points.

3.2.1 The A magnetization

At different sampling sites of the Haliburton intrusions, the high-T_B A component has T_B's near 580°C consistent with magnetite as the dominant magnetic carrier, and T_B's well above 600°C, implying that hematite carries the remanence.

Unmistakable evidence that the high-T_B A component is carried by both magnetite and hematite, even within individual samples, is found in the Magnetawan metasediments (McWilliams and Dunlop, 1975). Figure 3 shows a set of recently measured thermal demagnetization curves for Magnetawan samples that contain purely the A component. The closely spaced heating steps leave no doubt that although the bulk of the A blocking temperatures cluster near 580°C, the Curie point of magnetite, a significant fraction of the A magnetization has blocking temperatures

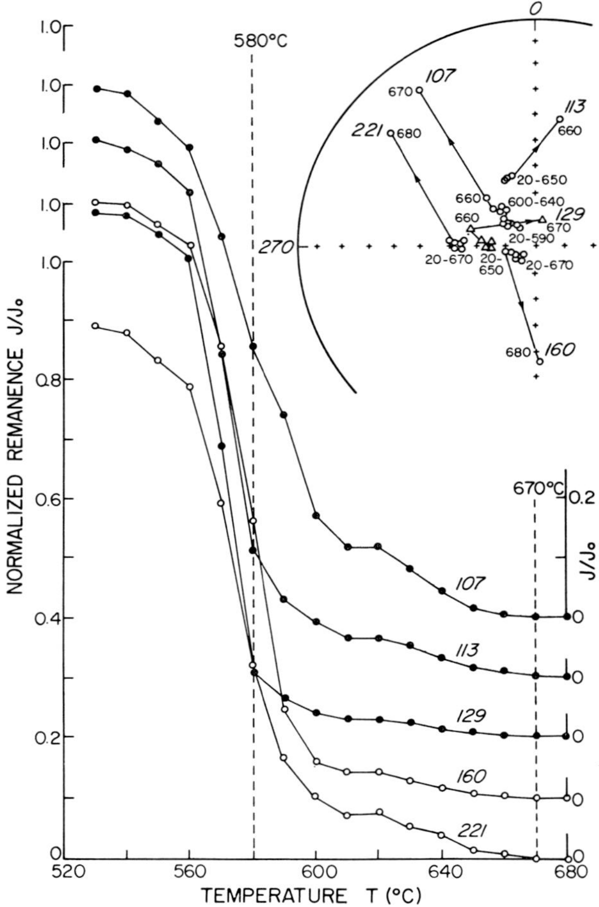

Fig. 3. Thermal demagnetization characteristics of typical A-remanence samples from the Magnetawan metasediments. The curves have been stacked to spread them. The Curie temperatures of magnetite (580°C) and hematite (670°C) are shown as dashed lines. All inclinations are upwards and all points represent thermal data. Sample numbers are in italics.

up to 650–670°C. In other words, magnetite and hematite, coexisting within the same hand-sample, both carry an *A* magnetization.

It is difficult to envisage a low-temperature chemical process that would generate *both* magnetite and hematite at the same time (the time recorded by the *A* geomagnetic field direction). It is possible that the hematite fraction of the *A* magnetization is a CRM, but the magnetite fraction, at least, must be a TRM. The *A* component has *true* blocking temperatures of 550–580°C and possibly higher ones as well.

3.2.2 The B magnetization

The Haliburton *B* magnetization contains two fractions (BUCHAN and DUNLOP, 1976). One fraction has T_B's>450°C. It is undoubtedly of chemical origin since its highest blocking temperatures frequently overlap those of the *A* component.

The other fraction has 100°C$<T_B<450$°C. Because of its similarity to artificial low-temperature partial TRM's produced in the same samples (BUCHAN, 1977b), we believe it is a low-temperature thermal overprint. IRVING *et al.* (1974) independently reached the same conclusion about the origin of the 'secondary' component of NRM in the Morin anorthosite.

The very low blocking temperatures of the Haliburton *B* component rule out any reheating subsequent to the acquisition of the *B*. The high-T_B *A* component necessarily predates the low-T_B *B* component. If the Thanet *A* and *B* components date the same overprinting events as the Haliburton *A* and *B* components respectively, the Thanet *B* magnetization is a chemical overprint, with high *apparent* blocking temperatures.

4. Techniques for Establishing Absolute Ages of Superimposed Magnetizations

Once it has been shown that one or more of the components in a multi-component study is of thermal origin, one can proceed to the second stage of the dating problem: estimating the absolute age of each thermal component.

First we date the body radiometrically, obtaining whole-rock and mineral-separate ages. For example, the ^{40}Ar/^{39}Ar stepwise heating technique gives two distinctly different ages for the Bark Lake (Haliburton) intrusion (Fig. 4): approximately 990 my for hornblende, and about 920 my for biotite. The reason for this apparent discrepancy is probably that these minerals became closed to Ar diffusion at different closure or blocking temperatures during the slow cooling that followed burial and metamorphism of the region.

Indeed, with a sufficient number of reliable age determinations from minerals with a wide range of closure temperatures, we can define an approximate temperature-time curve for the body. Magnetic blocking temperatures of the various partial TRM components can then be matched with this calibrated cooling curve to date the magnetizations.

Blocking temperatures in nature during slow cooling differ from blocking temperatures determined in rapid laboratory heatings. Hence, we must always take into

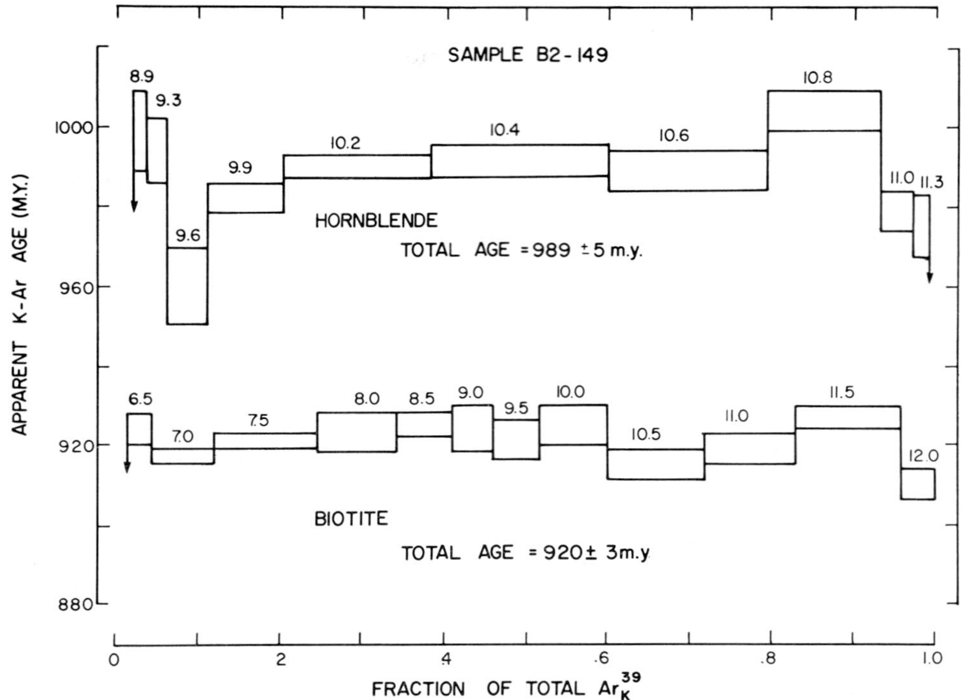

Fig. 4.   Hornblende and biotite ^{40}Ar/^{39}Ar age determinations for the Bark Lake diorite of the Haliburton instrusions. The length of the bars represents the volume fraction of Ar released at a given temperature (in hundreds of degrees Celsius). The height of the bars represents 1σ analytical error including all sources except error in the apparent age of the reference mineral and in the decay constants.

account the effects of a long heating-cooling history on both radiometric closure temperatures and magnetic blocking temperatures.

The time dependence of T_B is well known in the magnetic case. For example, Pullaiah *et al.* (1975) show that, according to the Néel (1949) theory of independent single-domain grains, a grain which is just unblocked at temperature T_{B1} in a laboratory-scale time t_1, will be unblocked in nature over a long time t_2 at a lower temperature T_{B2}:

$$\frac{T_{B2}}{J_S(T_{B2})H_C(T_{B2})} = \frac{\log(Ct_1)}{\log(Ct_2)}\frac{T_{B1}}{J_S(T_{B1})H_C(T_{B1})},$$

J_S being spontaneous magnetization, H_C microscopic coercive force, and C the characteristic frequency of thermal fluctuations ($\sim 10^9$ sec^{-1}). Laboratory blocking temperatures are readily extrapolated to geological blocking temperatures using this equation or its graphical representation. Type curves are given by Pullaiah *et al.* (1975) for Fe$_3$O$_4$ and αFe$_2$O$_3$ and by Dunlop and Hale (1977) for Fe$_{2.4}$Ti$_{0.6}$O$_4$.

Consider the partial TRM components of the Haliburton intrusions. We estimate that the A magnetization carried by magnetite was acquired at temperatures within about 10°C of those observed in the laboratory (550–580°C) while temperatures of no more than 250°C are necessary to explain $B\ T_B$'s of up to 400°C if the B remanence is carried by a low-titanium titanomagnetite.

The diffusion of argon from a rock-forming mineral also depends on a thermal activation mechanism. Accordingly, a grain from which a certain fraction of argon outgasses in a laboratory heating at temperature T_1 for a time t_1, will lose the same argon fraction in a geological time interval t_2 at a lower temperature T_2:

$$\frac{1}{T_2} = \frac{1}{T_1} - \frac{R}{E} \log\left(\frac{t_1}{t_2}\right),$$

R being the gas constant and E experimental activation energy.

Alternatively, the time-dependent history of a rock can be defined by considering that the sample was cooling continuously. DODSON (1973) derived an equation for radiometric blocking temperature:

$$(T_B)_{\text{radiometric}} = E/[R \log (A\tau D_0/a^2)],$$

a being a characteristic dimension for the grain, A a geometrical factor, D_0 the diffusion coefficient and τ a time constant related to cooling rate. In practice, the closure temperature is insensitive to uncertainties in τ (see DODSON, 1973), making this method promising for the study of Precambrian shields where cooling rate has been rather slow following burial metamorphism.

In an analogous fashion, YORK (1977a, b) is currently examining magnetic blocking temperatures in a continuously cooling orogen.

We have assumed a slow-cooling model (5°C/my near the blocking temperature) to interpret the Argon data from Bark Lake (Haliburton) rocks shown in Fig. 4, since the presence of well-defined plateaus in the coexisting biotite and hornblende probably restricts any reheating episode to temperatures less than 200–300°C (e.g., BERGER, 1975). We estimate the hornblende and biotite of Fig. 4 to have closed to Argon diffusion at 620 ± 40 (1σ) and about 225°C respectively. Unfortunately Dodson's formula could not be applied to the biotite data so we quote a geological estimate (TURNER and FORBES, 1976). We emphasize that these results are preliminary and details of the Argon work will be published separately.

The estimated closure temperatures of 620°C for hornblende and 225°C for biotite suggest that the hornblende and biotite ages of 990 and 920 my, likely bracket the acquisition times of the Haliburton A and B magnetizations.

5. Conclusions

Many of the techniques necessary for resolving metamorphically overprinted magnetizations and establishing their relative and absolute ages have yet to be fully developed. Nevertheless, as the case histories of Grenville rocks described in this paper demonstrate, these problems may be solved by a systematic investigation of the distribution of various magnetic components in different rock-forming minerals, the thermal and/or chemical origin of NRM, and the variation of radiometric and magnetic closure or blocking temperatures with time.

Support has been provided by the National Research Council of Canada to G.W. Berger as part of a Negotiated Development Grant in Earth Sciences at the University of Toronto and to D. York and D.J. Dunlop through operating grants.

REFERENCES

BERGER, G.W., ^{40}Ar/^{39}Ar step heating of thermally overprinted biotite, hornblende and potassium feldspar from Eldora, Colorado, *Earth Planet. Sci. Lett.*, **26**, 387–408, 1975.

BUCHAN, K.L., Magnetic overprinting in the Thanet gabbro complex, 1977a (in preparation).

BUCHAN, K.L., Rock magnetic and paleomagnetic studies of multi-component remanences in metamorphosed rocks of the Grenville Province of the Canadian Precambrian Shield, Ph. D. thesis, University of Toronto, Toronto, 1977b.

BUCHAN, K.L. and D.J. DUNLOP, Magnetization episodes and tectonics of the Grenville Province, *Nature Phys. Sci.*, **246**, 28–31, 1973.

BUCHAN, K.L. and D.J. DUNLOP, Paleomagnetism of the Haliburton intrusions: Superimposed magnetizations, metamorphism and tectonics in the late Precambrian, *J. Geophys. Res.*, **81**, 2951–2967, 1976.

DUNLOP, D.J. and C.J. HALE, Simulation of long-term changes in the magnetic signal of the oceanic crust, *Can. J. Earth Sci.*, **14**, 716–744, 1977.

DODSON, M.H., Closure temperature in cooling geochronological and petrological systems, *Contrib. Mineral. Petrol.*, **40**, 259–274, 1973.

FAHRIG, W.F., K.W. CHRISTIE, and E.J. SCHWARZ, Paleomagnetism of the Mealy Mountain anorthosite suite and of the Shabogamo gabbro, Labrador, Canada, *Can. J. Earth Sci.*, **11**, 18–29, 1974.

HALLS, H.C., A least-squares method to find a remanence direction from converging remagnetization circles, *Geophys. J. R. Astron. Soc.*, **45**, 297–304, 1976.

IRVING, E. and J.C. McGLYNN, Polyphase magnetization of the Big Spruce Complex, Northwest Territories, *Can. J. Earth Sci.*, **13**, 476–489, 1976.

IRVING, E., R.F. EMSLIE, and H. UENO, Upper Proterozoic paleomagnetic poles from Laurentia and the history of the Grenville structural province, *J. Geophys. Res.*, **79**, 5491–5502, 1974.

McWILLIAMS, M.O. and D.J. DUNLOP, Precambrian paleomagnetism: Magnetizations reset by the Grenville orogeny, *Science*, **190**, 269–272, 1975.

NÉEL, L., Théorie du traînage magnétique des ferromagnétiques en grains fin avec applications au terres cuites, *Ann. Géophys.*, **5**, 99–136, 1949.

PULLAIAH, G., E. IRVING, K.L. BUCHAN, and D.J. DUNLOP, Magnetization changes caused by burial and uplift, *Earth Planet. Sci. Lett.*, **28**, 133–143, 1975.

ROY, J.L. and P.L. LAPOINTE, The paleomagnetism of Huronian red beds and Nipissing diabase; Post-Huronian igneous events and apparent polar path for the interval −2300 Ma to −1500 Ma for Laurentia, *Can. J. Earth Sci.*, **13**, 749–773, 1976.

SCHWARZ, E.J., Depth of burial from remanent magnetization: the Sudbury irruptive at the time of diabase intrusion (1250 Ma), *Can. J. Earth Sci.*, **14**, 82–88, 1977.

TURNER, D.L. and R.B. FORBES, K-Ar studies in two deep basement drill holes: a new geological estimate of argon blocking temperature for biotite, *EOS, Trans. Am. Geophys. Union*, **57**(4), 353, 1976.

UENO, H., E. IRVING, and R.H. McNUTT, Paleomagnetism of the Whitestone anorthosite and diorite, the Grenville polar track, and relative motions of the Laurentian and Baltic shields, *Can. J. Earth Sci.*, **12**, 209–226, 1975.

VITORELLO, I. and R. VAN DER VOO, Late Hadrynian and Helikian pole positions from the Spokane Formation, Montana, *Can. J. Earth Sci.*, **14**, 67–73, 1977.

WU, Y.T., M. FULLER, and V.A. SCHMIDT, Microanalysis of N.R.M. in a granodiorite intrusion, *Earth Planet. Sci. Lett.*, **23**, 275–285, 1974.

YORK, D., A formula describing rock magnetic and isotopic blocking temperatures, *Earth Planet. Sci. Lett.*, 1977a (in press).

YORK, D., Magnetic Blocking Temperatures, *Earth Planet. Sci. Lett.*, 1977b (in press).

ZIJDERVELD, J.D.A., A.C. demagnetization of rocks: Analysis of results. In *Methods in Paleomagnetism*, edited by D.W. Collinson, K.M. Creer, and S.K. Runcorn, p. 254–286, Elsevier, Amsterdam, 1967.

Adv. Earth Planet. Sci., **1**, 179–187, 1977

Does TRM Occur in Oceanic Layer 2 Basalts?

J.M. HALL

Department of Geology, Dalhousie University,
Halifax, Nova Scotia, Canada

(Received June 20, 1977)

It is shown that, as a consequence of rapid low temperature oxidation, the magnetization of at least the upper 600 m of oceanic layer 2 almost everywhere, except immediately adjacent to ridge crests, must be regarded as being chemical remanent rather than thermoremanent in nature. While most magnetic properties are changed drastically as a result of this alteration process, it seems that initial TRM direction is unlikely to be altered by more than 10°, and then only when original magnetizing and dipole fields differ markedly in direction. Increased scatter in direction appears to be a more general consequence of CRM replacement of TRM.

1.   Introduction

The essence of the model of oceanic crust as a recorder of geomagnetic field reversals is in the assumption that field directions are reliably recorded by at least some ocean crust rocks for up to 200 my. This assumption can be made with conviction if fairly fine-grained, magnetite bearing rocks are involved, and the initial magnetization is of thermoremanent nature. The last stipulation follows from the well known strong and stable nature of laboratory TRM in the appropriate fine-grained basaltic rocks.

Serious criticism of the model follows if it can be shown that the magnetization of oceanic crustal rocks is not initial thermoremanence, but was acquired by some other process well after initial cooling. A delay in remagnetization of 10^5–10^6 yr. in the Neogene could be very serious since a stable direction opposing the initial TRM might be acquired. It is critical to test whether the NRM retains the original TRM direction.

The plan of this paper is first to identify the nature of the magnetization of ocean crust rocks and then to look carefully at the meaning, in terms of geomagnetic field record, of the remanence directions.

2.   Nature of Magnetization of Ocean Crust Rocks

The nature of the magnetization of upper layer 2 of the oceanic crust, which consists largely of pillow basalts, has become clear from a series of studies extending over the last 15 years. Results to 1970 are summarized in WATKINS and PASTER (1971) while more recent work is described in LOWRIE (1974, 1977). Presently information is available for basalts forming the upper surface of the layer from a wide range of geographic locations in all the major oceans, and to a depth of 500 to 600 m at a limited number of locations in the Atlantic. Basalts which are the least altered contain optically homogeneous iron-titanium oxide magnetic minerals, titanomagnetites, with composition close to $x=0.65$ in $x\mathrm{Fe_2TiO_4}(1-x)\mathrm{Fe_3O_4}$ (JOHNSON and HALL (in press)). However, in terms of distribution, these little altered basalts are rare, only occurring among the youngest ($\simeq 10^4$ yr.) pillow flows on ridge crests or in occasional older massive flows (JOHNSON and HALL, op. cit.). Most basalts of layer 2 show optical evidence of alteration, with a range from minor indications of oxidation to almost complete replacement of oxides by silicates. A convenient way to quantify the observed changes is by the degree of cation deficiency of the titanomagnetite. This is obtained, once an x value is known or assumed, from the Curie temperatures of basalt samples or the unit cell edge of separated oxides. Unoxidized or stoichiometric titanomagnetites from layer 2 basalts have Curie temperatures close to 120°C. This temperature implies an x value of close to 0.65 and a cation deficiency, z, of zero. As oxidation proceeds Curie temperature rises continuously to a maximum value of about 420°C, corresponding to a value of 1.00, or complete replacement or removal of $\mathrm{Fe^{2+}}$ ions from the titanomagnetite lattice. A number of studies describe the abundance of cation deficient forms compared with stoichiometric forms of titanomagnetite (OZIMA and OZIMA, 1971; LOWRIE et al., 1973a, b). Other studies show that titanomagnetites from basalts forming the upper surface of layer 2 are appreciably cation deficient within a few million years of formation (SCHAEFFER and SCHWARZ, 1970; JOHNSON and ATWATER, 1977) while RYALL et al. (1977) show that a high average state of cation deficiency extends to at least 560 m depth in layer 2 by at the most 3.5 my after crustal formation (Fig. 1).

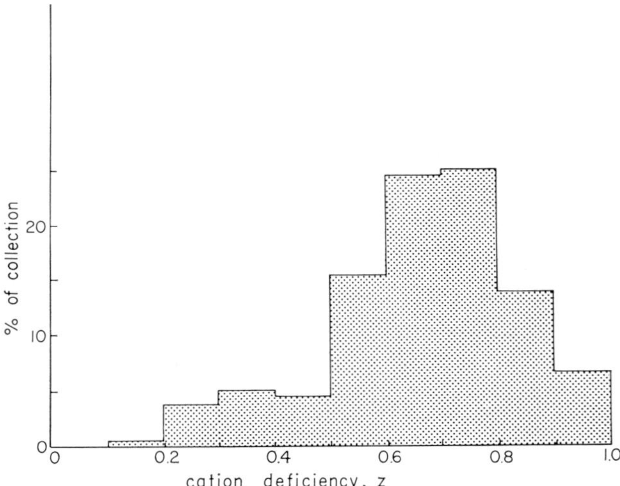

Fig. 1.   Distribution of titanomagnetite cation deficiency in 447 oceanic layer 2 basalts sampled during Deep Sea Drilling Project Leg 37. Age range 3.5 to 16 my. Depth range 0 to 560 m sub-basement (from RYALL et al. 1977).

The physical and chemical conditions in which cation deficiency is produced in layer 2 basalts seem clear. A combination of in situ temperature measurements, secondary mineral structures, oxygen isotope determinations and laboratory tests of the thermal stability of cation deficient titanomagnetites show that oxidation by sea water at not more than a few tens of degrees centigrade is the only common mechanism (YEATS et al., 1976; HYNDMAN et al., 1976; MUEHLENBACHS, 1977).

3.   Magnetic Changes Accompanying Low Temperature Oxidation

Figure 2 summarizes the evolution of a layer 2 basalt as a magnetic material. Optical mineralogy and thermomagnetic tests show that none or little of the original titanomagnetite remains in oxidized samples. For this reason, it seems necessary to regard the magnetization of the oxidized basalts as a chemical remanence (CRM) as suggested by IRVING et al. (1970) since the oxidation process takes place essentially isothermally at close to ocean bottom temperatures which are well below the Curie temperatures of the titanomagnetites. The presence of thermochemical remanence at depth in the crust is unproven but seems likely.

Figure 3 summarizes the major changes in magnetic properties that accompany increasing titanomagnetite cation deficiency. From the point of view of fidelity in preserving TRM directions we note that a very large decrease in remanence (NRM) intensity occurs during the alteration process. Study of the variation in NRM away from ridge crests, which is likely to be largely dependent on the progress of alteration, suggest that at least a factor of five overall decrease occurs over the full range of z (HALL, 1977).

It must be asked whether such major decrease in remanent intensity can occur without significant change in remanence direction. Several laboratory experiments led to the conclusion that remanence direction might persist unchanged during alteration (WILSON and SMITH, 1970; MARSHALL and COX, 1971). With increased

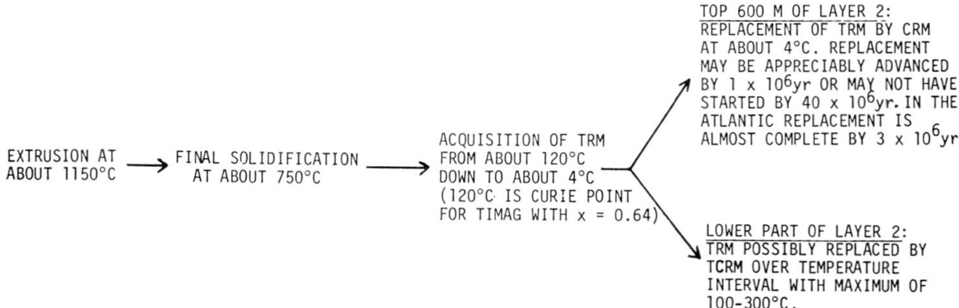

Fig. 2.   The life history of a submarine basalt as a magnetic material.

THE INFLUENCE OF LOW TEMPERATURE OXIDATION ON THE MAGNETIC PROPERTIES OF SUBMARINE BASALTS

PROPERTIES THAT <u>INCREASE</u>
IN VALUE AS OXIDATION
INCREASES:

CURIE TEMPERATURE: 120°C to 420°C for $z = 0$ to $z = 1$

MEAN DEMAGNETIZING FIELD (ALSO DEPENDENT ON GRAIN SIZE
VARIATION):   ABOUT AN ORDER OF MAGNITUDE FROM $z = 0$ to $z = 1$

PROPERTIES THAT <u>DECREASE</u>
IN VALUE AS OXIDATION
INCREASES:

NATURAL REMANENCE INTENSITY: ABOUT A FACTOR OF FIVE FROM $z = 0$
to $z = 1$

SATURATION MAGNETIZATION ⎫   BETWEEN A FACTOR OF FIVE AND AN
⎬   ORDER OF MAGNITUDE FROM $z = 0$ to
INITIAL SUSCEPTABILTIY   ⎭ $z = 1$

PROPERTY THAT SHOWS A PEAK
VALUE AT AN INTERMEDIATE
OXIDATION LEVEL

Q RATIO, PEAK AT ABOUT $z = 0.75$

Fig. 3.   The influence of low temperature oxidation on the magnetic properties of submarine
basalts.

knowledge of alteration mechanisms, and, in particular, discovery that the thermal
stability of cation deficient titanomagnetites was limited to temperatures below about
135°C (JOHNSON and MERRILL, 1973), the applicability of these experiments to natural conditions came into question.   Both studies described laboratory tests carried
ont at temperatures of up to 600°C.   We know now that these high temperatures
lead to phase splitting of cation deficient titanomagnetite (e.g., WAYMAN and EVANS,
1977), which is very rare in layer 2 basalts, and so do not represent alteration in
natural conditions.

A permissible way to proceed is to make no assumption from laboratory experiments about the possibility of directional change, but instead to look at the results of
measurements of collections of layer 2 basalts for evidence that bears on the question.
Relevant evidence occurs as alternating field cleaned NRM inclinations for sets of
samples from basaltic units forming upper layer 2 in the Atlantic and Pacific.   In
several instances, twenty or more semioriented (azimuthally unoriented) samples are
available from a single lithologic unit (YEATS et al., 1976; MELSON et al., 1977).   For
each set it can be safely assumed that initial magnetization took place sufficiently
quickly (1 to 100 yr.) and that all the samples in the set had the same initial TRM
directions.   Since systematic alternating field demagnetization has been carried out
for all samples, we can be sure that change in cleaned inclination from sample to
sample in a set is not the result of the presence of perturbing viscous magnetic components.   Finally, as a result of the uneven progress of oxidation in submarine basaltic
flows, many of the sets show quite a wide range in titanomagnetite cation deficiency.
Plots of the variation of stable inclination with z, the degree of cation deficiency
will show whether any directional change accompanies low temperature alteration.
Figures 4 and 5 illustrate the data.   In all cases the spread of inclinations, which is
typically close to 20°, in these large sets, yields well confined mean inclinations with

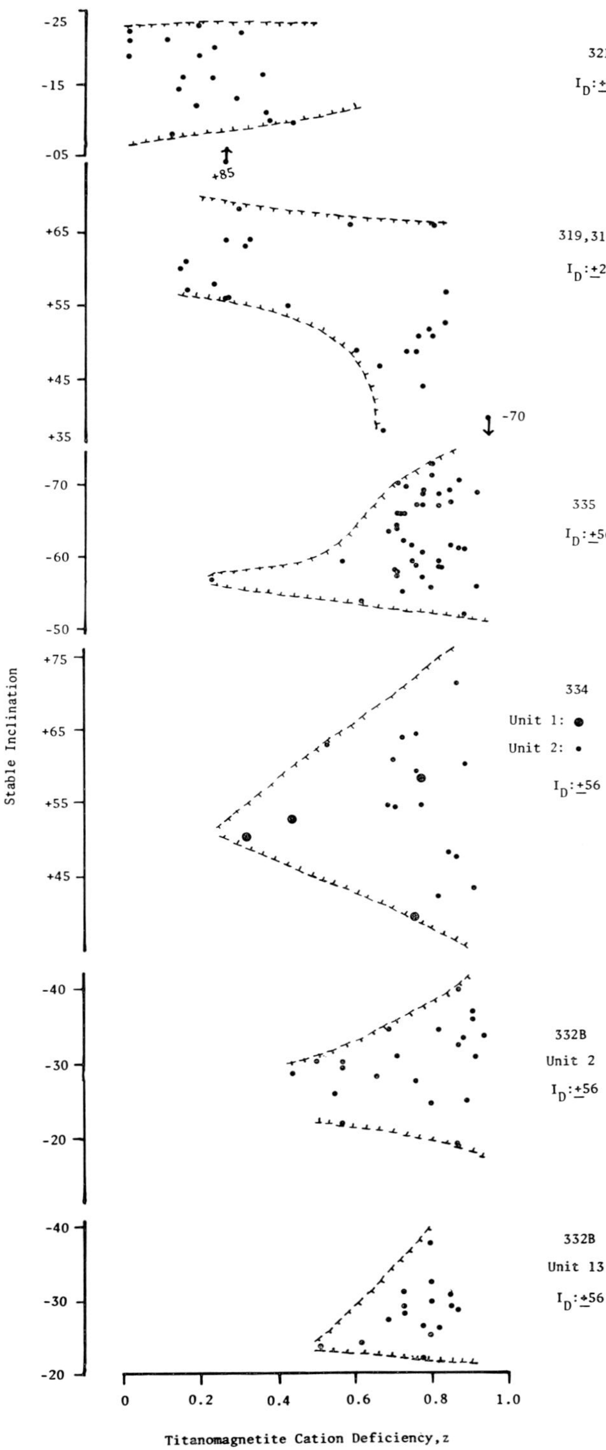

Fig. 4.   The variation of stable inclination with degree of titanomagnetic cation deficiency in six oceanic layer 2 basalt units. (The basic data is available in the Initial Reports of the Deep Sea Drilling Project, Leg 34 and 37: YEATS *et al.*, 1976 and MELSON *et al.*, 1977.)

Site 319

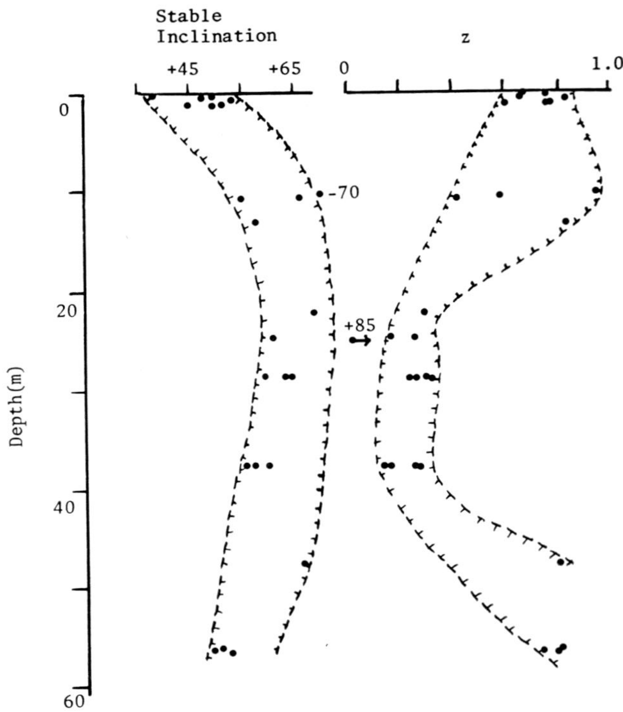

Fig. 5.   Variation of stable inclination and titanomagnetite cation deficiency
with depth in layer 2 basalts at D.S.D.P. Site 319 (from several sources in
YEATS *et al.*, 1976).

standard derivations of the mean value of from 1° to 4°. At our present level of understanding of paleomagnetic directions, these small uncertainties are interpreted as implying that original TRM inclinations are well preserved in the cleaned NRM inclinations. At the same time, there are strong hints in many of the distributions that cleaned inclinations do vary with z. Referring to Fig. 4 we note that:

1)   In five of six sets the scatter in inclinations increases with increasing z. The sixth set (321) contains only low z ($z \lesssim 0.4$) samples and in this respect differs from the other sets which are usually dominated by high z ($z \gtrsim 0.5$) samples.

2)   In two examples (319, 319A and 332B, Unit 2) the distribution of points is sufficient to suggest that a small change in average direction occurs as z increases. Approximately ten and six degrees average angular change, respectively, are apparent for the ranges in z involved. These figures are obtained by fitting a smooth curve to average values of stable inclination for cation deficiency intervals of 0.1. Similar cases might be made for 335 and 332B, Unit 13 but here the forms of the envelopes are determined by single relatively low z value samples.

The data for 319/319A is shown in a slightly different format in Fig. 5. It is clearly seen that the low z zone in the middle of the sequence corresponds to relatively high ($+55°$ to $+65°$) stable inclinations while high z zones above and below correspond to relatively low ($+45°$ to $+55°$) inclinations.

Before considering possible geophysical implications of these variations it is necessary to point out the deficiencies in the data. First, the quality of the data is barely adequate to distinguish the rather small angular changes which may occur as the result of alteration. A reasonable estimate of the average angular uncertainty in a data point is $\pm 5°$. Major contributions to this uncertainty are the frequent need to sample slightly barrel-shaped core samples, formed by the breakup and rotation of sections of core during drilling, rather than perfectly cylindrical sections, and in defining cleaned inclinations (HALL and RYALL, 1977). It is only because large numbers of samples are available in each set that trends may be identified at all. Secondly, only one example (319, 319A) contains samples covering most of the range in z (from $z \sim 0.2$ to ~ 0.8). The low z samples in this set are from massive basalts intercalated between pillow sequences. All the other sets are from pillow sequences and the rapid progress of alteration in lava pillows has left little or no low z material for study.

Explanation of the inclination trends with increasing z may be as follows:

1) Increased scatter in inclination for $z > 0.5$ may represent CRM acquisition during the higher degrees of alteration at different times within one polarity epoch or during polarity epochs of different senses. This seems likely to happen since, although alteration is rather quick, being largely completed in 2 to 3 million years, the length of polarity epochs, at least during the young Neogene is an order of magnitude less than these times (Cox, 1969).

It must be emphasized that the tendency for CRM to follow the direction or polarity of the field during acquisition is at best weak and usually the initial magnetizing field direction still dominates.

2) Change in average cleaned inclination with increasing z, which occurs in two examples in addition to increased scatter, may be related to marked differences between initial magnetizing and local dipole field inclinations. Thus, an atypically *steep* initial magnetizing field of between $+55°$ and $+65°$ inclination is $30°$ to $40°$ away from the normal dipole field inclination of $+25°$ for the site of 319/319A. Again, for 332B, Unit 2, atypically *shallow* initial magnetizing field inclination of about $+25°$ is $30°$ away from the normal dipole inclination of $+56°$. For both examples the effect of CRM acquisition has been to rotate stable inclinations slightly towards normal dipole field inclinations. Presumably this effect represents in both cases a *major* part of CRM acquired in *normal* polarity dipole fields. For the originally reversely magnetized Unit 2 of Hole 332B, of age 3.5 my, major alteration-induced magnetic changes took place during the succeeding Gauss normal polarity epoch and this is likely to be the period during which bias in CRM directions towards the normal dipole field occurred. The combined circumstances of non-dipolar TRM inclinations and CRM acquisition during a single polarity epoch have been proposed independently by MERRILL (1975) as optimal for CRM and TRM inclinations to differ. The contribution of the present study is to demonstrate the existence and magnitude of the effect. If tectonic tilting is responsible for the observed shallow stable incli-

nations for Unit 2 of Hole 332B, then the time of tilting can be determined from this magnetization model as being between 3.5 and 3.3 mybp, the beginning of the Gauss normal epoch. In terms of the present median valley in the FAMOUS area (MACDONALD and LUYENDYK, 1977) tilting would have taken place at the walls of the inner rift valley. The age of the 319-319A set is not sufficiently well known to suggest a particular normal polarity epoch for CRM acquisition.

4.   Conclusions

It is shown that CRM is the dominant type of magnetization of upper oceanic layer 2 basalts. This is in contrast with earlier suggestions that TRM was the dominant form of magnetization. However, regardless of the widespread occurrence of CRM, original cooling TRM directions are generally well preserved, as is required in the current model for linear magnetic anomaly generation. Small ($\leq 10°$) changes in inclinations are likely to occur during CRM acquisition and the sense and magnitude of changes appear to depend on the relationship of the original magnetization to the reversal history during the period of alteration.

I would like to thank Mary Ann Annand and Margaret Odell for typing and Alice Scott for drafting. The research leading to this paper was supported by National Research Council of Canada grants A-7812, T-0475 and D-53.

REFERENCES

COX, A., Geomagnetic reversals, *Science*, **163**, 237–245, 1969.

HALL, J.M., Age variation in the magnetic properties of oceanic crust, *EOS, Trans. Am. Geophys. Union*, **58**, 377–378, 1977.

HALL, J.M. and P.J.C. RYALL, Paleomagnetism of basement rocks, Leg 37. in *Initial Reports of the Deep Sea Drilling Project Leg 37.*, edited by W. Melson and F. Aumento *et al.*, U. S. Government Printing Office, Washington, D.C., 1977.

HYNDMAN, R.D., R.P. VON HERZEN, A.J. ERICKSON, and J. JOLIVET, Heat flow measurements in deep crustal holes on the Mid-Atlantic Ridge, *J. Geophys. Res.*, **81**, 4053–4060, 1976.

IRVING, E., J.K. PARK, S.E. HAGGERTY, F. AUMENTO, and B.D. LONCAREVIC, Magnetism and opaque mineralogy of basalts from the Mid-Atlantic Ridge at 45°N, *Nature*, **228**, 974–976, 1970.

JOHNSON, H.P. and T. ATWATER, Magnetic study of basalts from the Mid-Atlantic Ridge, Lat. 37°N, *Bull. Geol. Soc. Am.*, **88**, 637–647, 1977.

JOHNSON, H.P. and J.M. HALL, A detailed rock magnetic and opaque mineralogy study of basalts from the Nazca Plate, *Geophys. J. R. Astron. Soc.*, (in press).

JOHNSON, H.P. and R.T. MERRILL, Low temperature oxidation of a titanomagnetite and the implications for paleomagnetism, *J. Geophys. Res.*, **78**, 4938–4949, 1973.

LOWRIE, W.R., Oceanic basalt magnetic properties and the Vine and Matthews hypothesis, *Geophys. J.*, **40**, 513–536, 1974.

LOWRIE, W.R., Intensity and directions of magnetization in oceanic basalts, *J. Geol. Soc., London*, **133**, 61–82, 1977.

LOWRIE, W.R., R. LOVLIE, and N.D. OPDYKE, Magnetic properties of Deep Sea Drilling Project basalts from the north Pacific Ocean, *J. Geophys. Res.*, **78**, 7647–7660, 1973a.

LOWRIE, W.R., R. LOVLIE, and N.D. OPDYKE, The magnetic properties of Deep Sea Drilling Project basalts from the Atlantic Ocean, *Earth Planet. Sci. Lett.*, **17** 338–349, 1973b.

MACDONALD, K.C. and B.P. LUYENDYK, Deep-tow studies of the structure of the Mid-Atlantic Ridge crest near latitude 37°N, *Bull. Geol. Soc. Am.*, **88**, 621–636, 1977.

MARSHALL, M. and A. COX, Effect of oxidation on the natural remanent magnetization of titano-magnetite in suboceanic basalt, *Nature*, **230**, 28–31, 1971.

MELSON, W., F. AUMENTO et al., *Initial Reports of the Deep Sea Drilling Project*, Vol. 37, U. S. Government Printing Office, Washington D.C., 1977.

MERRILL, R.T., Magnetic effects associated with chemical changes in igneous rocks, *Geophys. Surv.*, **2**, 277–311, 1975.

MUEHLENBACHS, K., Oxygen isotope geochemistry of rocks from the Deep Sea Drilling Project Leg 37, *Can. J. Earth Sci.*, **14**, 771–776, 1977.

OZIMA, M. and M. OZIMA, Characteristic thermomagnetic curve in submarine basalts, *J. Geophys. Res.*, **76**, 2051–2056, 1971.

RYALL, P.J.C., J.M. HALL, J. CLARK, and T. MILLIGAN, Magnetization of oceanic crustal layer 2—results and thoughts after Deep Sea Drilling Project Leg 37, *Can. J. Earth Sci.*, **14**, 684–706, 1977.

SCHAEFFER, R.M. and E.J. SCHWARZ, The Mid-Atlantic Ridge near 45°N, IX. Thermomagnetics of dredged samples of igneous rocks, *Can. J. Earth Sci.*, **7**, 268–273, 1970.

WAYMAN, M.L. and M.E. EVANS, Oxide microstructures and the magnetic properties of Leg 37 basalts, *Can. J. Earth Sci.*, **14**, 656–663, 1977.

WATKINS, N.D. and T.P. PASTER, The magnetic properties of igneous rocks from the ocean floor, *Philos. Trans. R. Soc., London A*, **268**, 507–550, 1971.

WILSON, R.L. and P.J. SMITH, The nature of secondary natural magnetizations in some igneous and baked rocks, *J. Geomag. Geoelectr.*, **20**, 367–380, 1970.

YEATS, R.S., S.R. HART et al., *Initial Reports of the Deep Sea Drilling Project*, Vol. 34, U. S. Government Printing Office, Washington D.C., 1976.

Adv. Earth Planet. Sci., **1**, 189–207, 1977

The Effects of Alteration on the Natural Remanent Magnetization of Three Ophiolite Complexes: Possible Implications for the Oceanic Crust

Shaul Levi* and Subir K. Banerjee

Department of Geology and Geophysics, University of Minnesota, Minneapolis, Minnesota, U.S.A.

(Received July 6, 1977)

Magnetic properties are compared for the pillow basalts and sheeted dike complexes for the following three ophiolite suites: Macquarie Island ($\sim$27 million years), Troodos Massif, Cyprus ($\sim$middle Cretaceous), and Smartville complex, California ($\sim$Jurassic). The magnetic properties of the pillow basalts strongly depend on their degree of alteration. From the least altered, low-temperature-oxidized, upper pillow basalts of Macquarie Island and the pillow basalts of the Troodos Massif (zeolite facies metamorphism) to the most altered pillow basalts of the Smartville complex (greenschist facies metamorphism) the average intensity of the natural remanent magnetization (NRM) decreases from a maximum of 80×10^{-4} to 0.50×10^{-4} gauss; the Koenigsberger ratio (Q) decreases from 6 to 0.3; the contribution due to viscous remanent magnetization (VRM) increases, and the NRM stability with respect to alternating fields decreases. For the most altered pillow basalts of the Smartville complex the NRM is probably predominantly a VRM, acquired during the present polarity epoch. Thermomagnetic analyses for the least altered pillow basalts exhibit the characteristic behavior of low-temperature-oxidized cation deficient titanomagnetites, while the more altered pillow basalts have reversible thermomagnetic curves with magnetite-like Curie points, probably caused in situ by thermally induced unmixing of the low-temperature-oxidized titanomagnetites. The sheeted dike complexes have NRM intensities of about 10×10^{-4} gauss and Q values of about 0.8, and most of their magnetic properties do not vary much between the three ophiolite complexes. All the sheeted dike samples have reversible thermomagnetic curves with magnetite-like Curie points. If these ophiolite complexes represent material which was originally formed at spreading centers and if their alteration represents ocean floor metamorphism, then with progressive alteration the pillow basalts lose their magnetic recording qualities and become poorer sources for the marine magnetic anomalies, and the underlying sheeted dike complexes become relatively more important. The most probable site for the occurrence of the sea floor metamorphism is in the relatively high temperature, high heat flow regions near spreading centers.

1. Introduction

Understanding the source layer of the marine magnetic anomalies has been the

---

* Now at School of Oceanography, Oregon State University, Corvallis, Oregon, U.S.A.

focus of a growing number of paleomagnetic and rock magnetic studies. The magnetic properties of unoriented and partly oriented (inclination only) basalts dredged and cored in young oceanic crust near the mid-ocean spreading centers (IRVING *et al.*, 1970; CARMICHAEL, 1970; MARSHALL and COX, 1971a; JOHNSON and ATWATER, 1977) have been interpreted to suggest that only a relatively thin magnetized layer (<1 km) is responsible for the observed anomalies. An approximately 500 m thick magnetic layer has also been obtained from the inversion of magnetic anomalies and from the frequent correlation between magnetic and topographic features (TALWANI *et al.*, 1971; SCLATER and KLITGORD, 1973; ATWATER and MUDIE, 1973). However, recent data from the Deep Sea Drilling Project (DSDP), with maximum basement penetration of 583 m, suggest that a magnetic layer several kilometers thick is required to account for the anomalies (SCIENTIFIC PARTY DSDP, LEG 37, 1975; JOHNSON, 1976); this conclusion is inferred because: 1) the NRM intensities are relatively low; 2) apparent reversals are present in the cored columns; 3) a significant fraction of the cored material consists of relatively non-magnetic sediment, rubble, breccia or voids. Thus, DSDP has provided the first hard data suggesting that even in oceanic crust as young as 7 million years (Legs 45 and 46) a 500 m thick basement layer is probably not sufficient to account for the observed anomalies and that the present drilling capabilities of the Glomar Challenger might not be sufficient to sample the entire thickness of the magnetic layer.

The intensity of NRM (natural remanent magnetization) of extrusive oceanic basalts has been shown to decrease dramatically within a few tens of kilometers away from the spreading centers (IRVING *et al.*, 1970; CARMICHAEL, 1970; JOHNSON and ATWATER, 1977) in general agreement with the comparable decrease of the amplitude of the magnetic anomalies. Low temperature oxidation of the remanence-carrying titanomagnetite minerals is now thought to be largely responsible for the observed decrease of NRM intensity (IRVING *et al.*, 1970; CARMICHAEL, 1970; OZIMA and OZIMA, 1971; MARSHALL and COX, 1971b; JOHNSON and MERRILL, 1972). BLAKELY (1976) analyzed magnetic anomaly data and found that the widths of the transition zones between oppositely magnetized oceanic crustal blocks increase with age. BLAKELY (1976) attributes this to thickening with age of the magnetic source layer enhanced by the magnetic degradation of upper extrusives. According to this model, the remanence of the extrusive basalts becomes relatively less important as they become older, and deeper rocks, presumably basaltic or doleritic dikes, become relatively more important as sources for the marine magnetic anomalies.

Ophiolite complexes, which are commonly thought to represent old oceanic crust, are the only presently available source for studying the deeper levels of the oceanic crust in relation to their contributions to the marine magnetic anomalies. In this paper we compare the magnetic properties of three ophiolites, restricting our discussion to the pillow basalts and the underlying dikes, which together probably account for much of the oceanic magnetic signal. The three ophiolites are: 1) the Macquarie Island ophiolite complex (54°S, 159°E), which is about 27 m.y. old (WILLIAMSON, 1974); 2) the Troodes Massif of Cyprus (35°N, 33°E), probably of middle Cretaceous age (MOORES and VINE, 1971); 3) the Smartville ophiolite complex of California (39°N, 121°W), tentatively of Jurassic age (SCHWEICKERT and COWAN, 1975). We find that in each of these ophiolite complexes the dikes contribute a significant and sometimes dominant fraction of the magnetic signal; however, the relative importance of the dike contribution is not related simply to the age of the ophiolite suite, but it depends more on the degree of alteration experienced by the pillow basalts of each complex.

2. Some Aspects of Metamorphism and Geochemistry

2.1 Macquarie Island

The doleritic dike swarms have been metamorphosed at different grades within the amphibolite facies (VARNE and RUBENACH, 1972). Some of the pillow basalts have undergone metamorphism at the greenschist facies level (designated as lower pillows); however, some of the pillows (upper pillows) are probably unmetamorphosed, although the presence of clay minerals suggests weathering of these rocks or at most very low grade metamorphism (VARNE and RUBENACH, 1972). CANN (1970) showed that the abundances of Ti, Y, Zr in ocean-floor basalts are affected little, if at all, by secondary weathering and metamorphism. PEARCE and CANN (1971) used these elements to discriminate between different volcanic regimes, and the method has been recently supported by studies of progressive metamorphism within a single basalt flow (SMITH and SMITH, 1976). Ti, Y, Zr analyses of Macquarie Island rocks show that they most closely resemble ocean-floor basalts (VARNE and RUBENACH, 1972). This conclusion has been confirmed more quantitatively by recent Sr isotope studies of WILLIAMS and MURTHY (1977).

2.2 Troodos Massif

The sheeted intrusive complex consists of basaltic dikes metamorphosed at the greenschist facies level (GASS and SMEWING, 1973; VINE and MOORES, 1972). The sheeted dikes grade upwards into the lower pillow lava complex, representing mostly zeolite facies metamorphism (GASS and SMEWING, 1973). Although the zeolite facies—greenschist facies metamorphic boundary is near the bottom of the pillow complex it does not coincide with the lithologic boundary, so that the 'metamorphic contact can occur either within the lower pillow lavas or the sheeted intrusive complex' (GASS and SMEWING, 1973). PEARCE and CANN (1971) analyzed the trace elements of 12 samples from the sheeted intrusive complex and pillow lavas of the Troodos Massif. When plotted on the Ti, Zr, Y triangle, these samples show close affinity to the ocean-floor basalts.

2.3 Smartville ophiolite complex

Preliminary petrological studies show that both in the sheeted dikes and pillow basalts epidote and amphibole are ubiquitous, and sphene and calcite are also frequently present. Thus, assuming a low P high T metamorphic evironment these mineral assemblages suggest metamorphism at greenschist facies grade. The data available thus far are consistent with similar metamorphic grade for both extrusive basalts and sheeted dike complex. In addition, the lack of schistosity and partial preservation of the igneous texture suggest that metamorphism was not accompanied by deformation.

For the ophiolite suites under discussion, the pillow lavas of the Troodos Massif and the upper pillows of Macquarie Island are the least metamorphosed, whereas the lower pillows of Macquarie Island and the pillow basalts from the Smartville complex have been more extensively metamorphosed at the greenschist facies level; at this point it is not possible to distinguish between the metamorphic grades of the lower pillow basalts of Macquarie Island and those of Smartville complex. Of the sheeted dike complexes, the one on Macquarie Island is most severely metamorphosed at the amphibolite facies level, and the dikes of the other two ophiolite suites represent greenschist facies metamorphism.

3.   Magnetic Properties

3.1   Thermomagnetic analyses

The magnetic phase of fresh extrusive basalts erupting at the mid-ocean ridges is stoichiometric titanomagnetite, whose ulvöspinel content is often near 60 mole percent (JOHNSON and HALL, 1977). At the temperature of the ocean floor these titanomagnetite particles are highly susceptible to low temperature oxidation to cation deficient spinels, titanomaghemites, which are metastable near room temperature, but which, even at temperatures as low as 150°C in some circumstances (JOHNSON and MERRILL, 1973; RYALL and ADE-HALL, 1975), unmix to titanium-poor titanomagnetities and relatively non-magnetic, titanium-rich phases. The various degrees of titanomagnetite alterations have characteristic thermomagnetic signatures (SCHAEFFER and SCHWARZ, 1970; OZIMA and OZIMA, 1971; OZIMA et al.,1974).

Figure 1 shows results of thermomagnetic analyses of samples from the three ophiolite complexes; and they are listed in order of increasing inferred age. Thermomagnetic experiments were done using a vibrating sample magnetometer in an external field of 2.5 k Oe and partial pressures of air less than 5×10^{-6} torr. Thermomagnetic analyses of the Macquarie Island and Smartville complexes were done with whole-rock fragments; however, magnetic separates were used in the runs with samples from the Troodos Massif, hence their higher values of saturation magnetization in Fig. 1.

The upper pillows of Macquarie Island exhibit irreversible thermomagnetic curves (BUTLER et al., 1976), consistent with their containing titanomagnetites which had been subjected to low temperature oxidation on the sea floor (SCHAEFFER and SCHWARZ, 1970; OZIMA and OZIMA, 1971). The lower pillows which had been metamorphosed at the greenschist facies grade exhibit reversible thermomagnetic curves, with Curie points very near 580°C: this behavior is consistent with a higher temperature event (prior to the thermomagnetic analysis) which caused unmixing of the

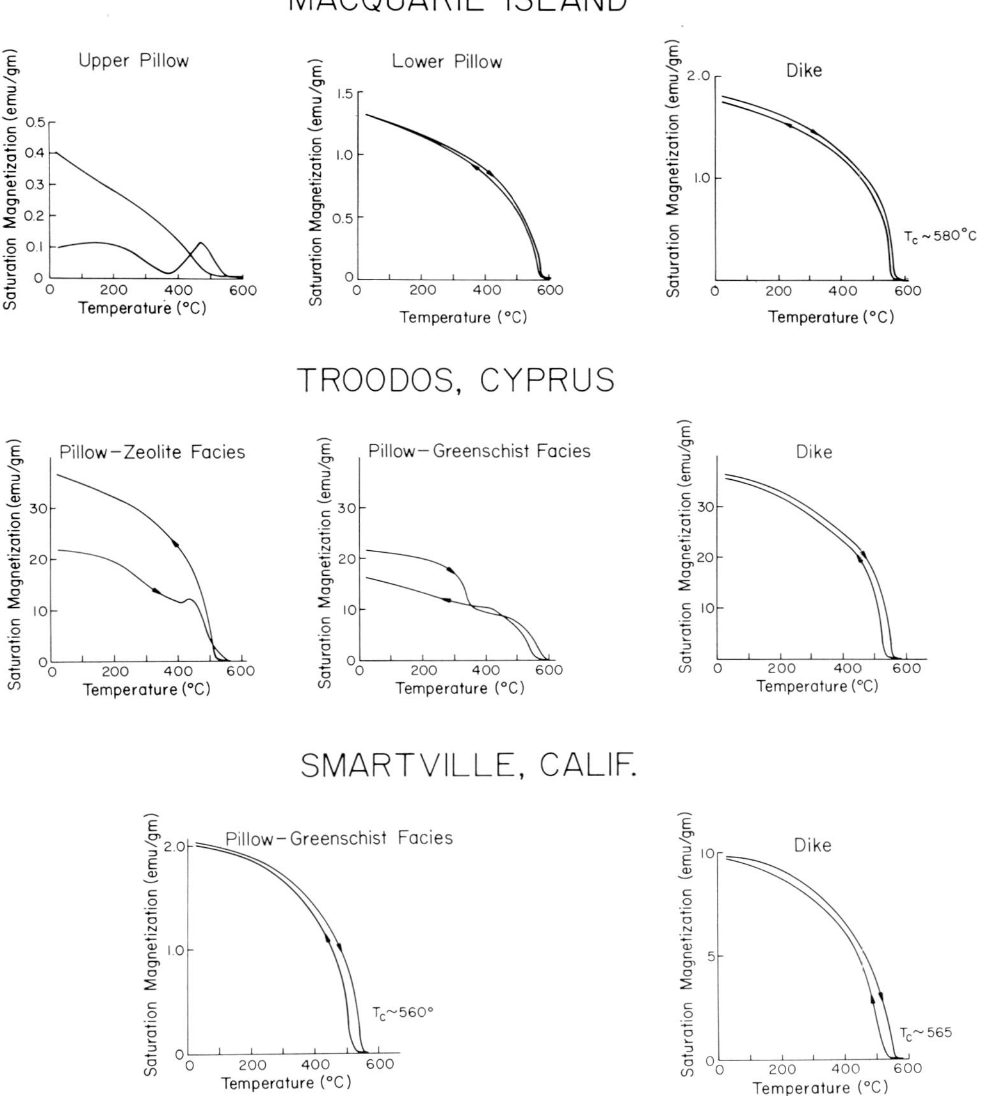

Fig. 1. Thermomagnetic curve for typical pillow basalt and dike samples for the three ophiolite complexes. The experiments were done in vacuum ($<5\times10^{-6}$ torr) and an applied field of 2.5×10^3 Oe.

titanomaghemite to nearly stoichiometric magnetite and additional non-magnetite phase (BUTLER *et al.*, 1976).

The Troodos Massif pillow basalts, which had undergone zeolite facies metamorphism exhibit irreversible thermomagnetic curves, consistent with the presence of low-temperature-oxidized titanomagnetite. The pillow basalt sample which had been metamorphosed to greenschist facies also has an irreversible thermomagnetic curve

MACQUARIE ISLAND
(a)
(b)
Pillows
Dikes
Number of Samples
Log₁₀ J_N
Log₁₀ Q
TROODOS , CYPRUS
(c)
(d)
Number of Samples
Log₁₀ J_N
Log₁₀ Q
SMARTVILLE , CALIF
(e)
(f)
Number of Samples
Log₁₀ J_N
Log₁₀ Q

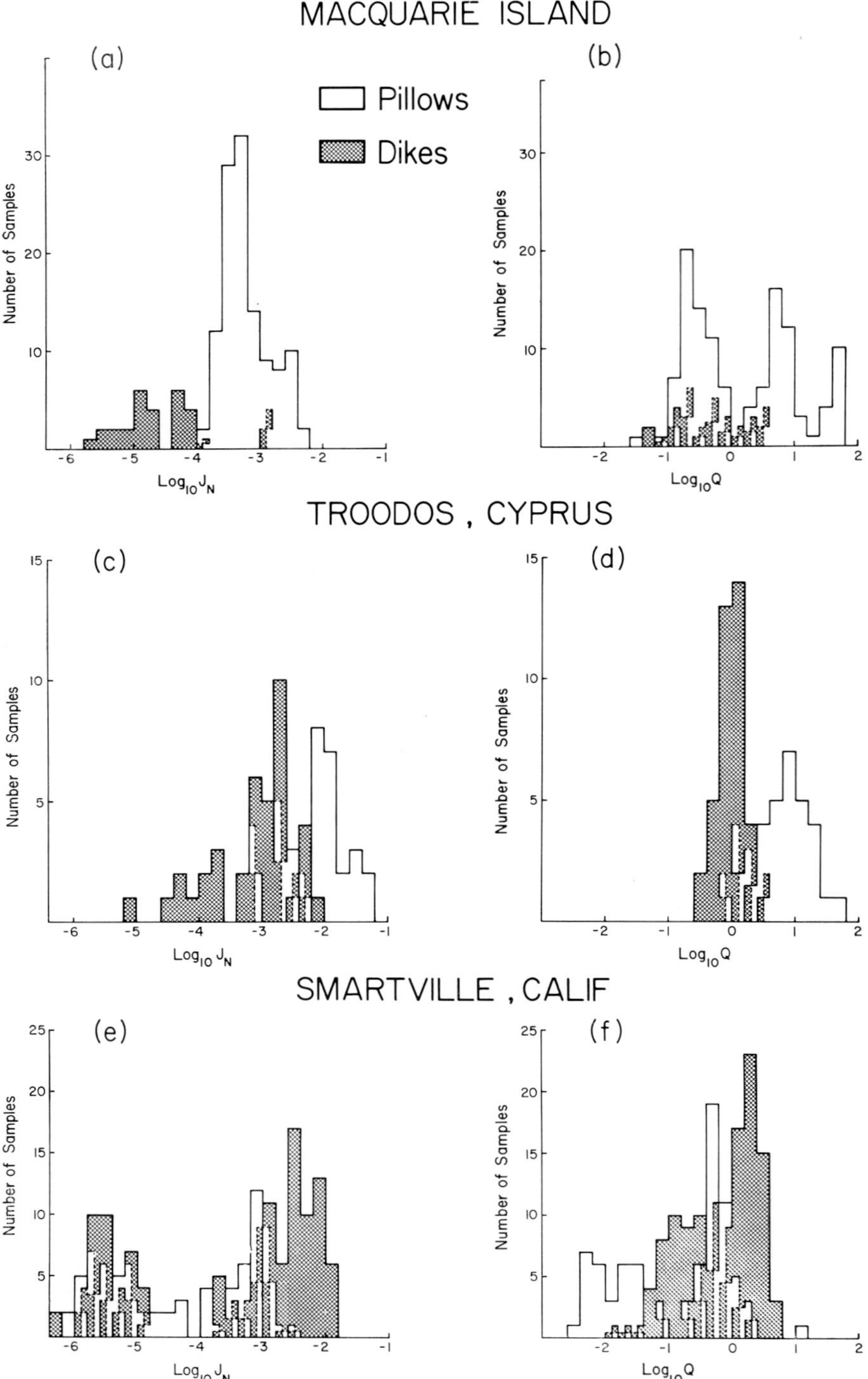

Fig. 2

but of a different kind. A substantial drop in J_S (saturation magnetization) occurs near 350°C and the remaining magnetization has a magnetite-like Curie point. J_S after cooling is considerably lower than its value before the thermomagnetic run. This behavior is analogous to that of the lower pillows of Macquarie Island. The metamorphic event presumably caused the unmixing of the low-temperature-oxidized titanomagnetite to highly magnetic, Ti-poor magnetite and relatively non-magnetic, Ti-rich phases. Subsequent to obduction, the magnetite experienced further subaerial low temperature weathering (oxidation) resulting in the production of maghemite, which being unstable to heating, converted to hematite near 350°C during the thermomagnetic run.

The pillow basalts of the Smartville complex have all been metamorphosed to greenschist facies, and they all give rise to reversible thermomagnetic curves with single Curie points near 580°C.

The thermomagnetic results for the pillow basalts of the three ophiolite complexes are therefore consistent with petrological data, suggesting that in the order of increasing degree of alteration they are: 1) The low-temperature-oxidized upper pillow basalts of Macquarie Island and the zeolite facies pillow basalts of the Troodos Massif and 2) the lower (greeschist facies) pillow basalts of Macquarie island and the greenschist facies pillows of the Troodos Massif and of the Smartville complex.

The intrusive sheeted dike complexes have all been metamorphosed at or beyond greenschist facies and, without exception, their thermomagnetic curves are reversible with a single magnetite-like Curie point.

3.2   Natural remanence and Koenigsberger ratio

Figure 2 shows histograms of J_N (gauss=emu/cm^3), the intensity of NRM of the three ophiolite complexes and corresponding histograms for the Koenigsberger ratio, Q, ratio of natural remanent to induced magnetization ($Q=J_N/\chi h$, where χ is the initial susceptibility and h is the inducing magnetic field). The ophiolite suites are listed in order of increasing inferred age. The data for the pillow basalts and sheeted complex of the Troodos Massif is taken directly from VINE and MOORES (1972; their Fig. 2), for which each value plotted represents the average of up to four specimens from each sample. For the Macquarie Island and Smartville complexes each point represents a single specimen. The data for the Macquarie Island pillow basalts is identical to that of BUTLER *et al.* (1976) and differences between Fig. 2 and Butler *et al.*'s Figs. 2 and 7 are due to different plotting-intervals of the logarithms of J_N and Q. The plotting-interval of Fig. 2 was determined by that of Fig. 2 of VINE and MOORES (1972). The logarithmic abscissa is used more for compactness than out of philosophical conviction. The average values of Table 1 represent the antilogarithms of the mid-point of the plotting-interval which contains

Fig. 2.   Histograms of NRM intensity (J_N) and Koenigsberger ratio (Q) of the pillow basalts (blank) and dike complexes (cross-hatched) of the three ophiolite complexes. In the determination of Q, h=0.65, 0.45, 0.50 Oe for the Macquarie Island ophiolite complex, Troodos Massif and Smartville ophiolite complex, respectively.

Table 1. Average values of NRM intensity ($\bar{J}_N$) and Koenigsberger ratio ($\bar{Q}$).

	$\bar{J}_N$ (gauss)	$\bar{Q}$		Degree of alteration
Pillow basalts				
Macquarie Island	5×10^{-4}	8	(10)	Upper pillows: low temperature oxidation
		0.3	(0.4)	Lower pillows: greenschist facies
Troodos Massif, Cyprus	80×10^{-4}	6.3	(5.7)	Zeolite facies mostly
Smartville complex, Calif.	0.5×10^{-4}	0.3	(0.3)	Greenschist facies
Sheeted dikes				
Macquarie Island	0.2×10^{-4}	0.5	(0.65)	Amphibolite facies
Troodos Massif, Cyprus	13×10^{-4}	1	(0.9)	Greenschist facies
Smartville complex, Calif.	13×10^{-4}	0.8	(0.8)	Greenschist facies

the median value. This procedure was adopted because of the incomplete availability of the Cyprus data. The values obtained in this manner can sometimes deviate from the true median value (e.g., determined for the Macquarie Island and Smartville complexes) by as much as 20 per cent, but these deviations are relatively unimportant for our purposes; we think that maintaining the consistency in the determination is more important in this case. Because of their different geographical locations, the intensity of the magnetic field (h) used in determining Q in Fig. 2 is different for the three complexes. $h = 0.65, 0.45, 0.50$ Oe for the Macquarie Island ophiolite, Troodos Massif and Smartville complex, respectively, and $\bar{Q}$ corresponding to these fields is listed in Table 1. However, because we are interested in the magnetic properties of these rocks independent of their present location, $\bar{Q}$ was adjusted for $h = 0.5$ Oe and listed in parentheses in Table 1.

It is evident from Fig. 2 and Table 1 that, relative to the ophiolite complexes that are being compared, the pillow basalts from the Troodos Massif are potentially the best magnetic recorders, both in terms of their high NRM intensity and their high Q value, showing the dominance of the remanent over the induced component. The NRM intensity of the Macquarie Island pillow basalts is on the average about one order of magnitude below the Cyprus pillows and there is, surprisingly, no significant difference in the NRM intensity of the greenschist facies and unmetamorphosed pillows (Butler *et al.*, 1976). However, $\bar{Q}$ of the unmetamorphosed pillows is more than 25 times greater than that of the greenschist facies pillow specimens. The average NRM intensity of the pillow basalts from the Smartville complex is about an order of magnitude lower than those from Macquarie Island and their $\bar{Q}$ value is similar to that of the similarly metamorphosed pillows from Macquarie Island. Since many factors affect the intensity of NRM, these values can obviously not be used by themselves to understand the cause for these variations. However, in conjunction with the $\bar{Q}$ values the J_N data suggest that the recording qualities of the pillow basalts deteriorate with increased alteration.

Despite the fact that for the sheeted dikes the NRM intensity is minimum for the most metamorphosed dike complex (Macquarie Island), the differences in the $\bar{Q}$

values for the different sheeted units are small and not significant. The picture that emerges from these data is that with increased alteration (not age) the underlying sheeted dikes become relatively more important as magnetic recorders because of the degradation of the magnetic recording quality of the pillow basalts. For the Smartville complex, in fact, the dikes are the predominant magnetic sources both in terms of $\bar{J}_N$ and $\bar{Q}$.

3.3   Viscous remanence

LOWRIE (1973) showed that VRM (viscouse remanent magnetization) contributes substantially to the NRM of some samples from layer 2 of the oceanic lithosphere. VRM acquisition experiments were conducted by us on samples from all three ophiolite complexes. The samples were first demagnetized by alternating fields (AF) to 1,000 Oe, which was taken as the base level for the VRM. VRM was produced in in a 0.45 Oe field and was monitored, typically for about ten weeks ($\sim 10^5$ minutes), and the VRM acquisition versus $\log_{10} t$ (minutes) data for representative samples are shown in Fig. 3. The VRM growth of most of the samples which were studied is usually linear on the semi-logarithmic graph to about 10^3 minutes; beyond this point, the rate of growth of VRM increases. This behavior is not uncommon in the literature (e.g., review paper on VRM by DUNLOP, 1973). To assess the importance of VRM acquired during the entire Brunhes normal polarity epoch the data obtained in a ten-week experiment is extrapolated to 7×10^5 years. Such gross extrapolations are always subject to large uncertainties, and this is particularly true in this case where the VRM versus $\log_{10} t$ curves are non-linear. In estimating the VRM produced during the last 700,000 years, we use a linear extraporation through the last two points of the VRM acquisition curves. Because of the increasing rate of change of acquired VRM, our projected Brunhes epoch VRM estimates probably represent a lower limit. The disadvantage of extrapolating VRM experiments done in the laboratory to geological time is that the more stable magnetic grains, having long relaxation times, are not affected during the short laboratory experiments. VRM experiments at progressively higher temperature would 'see' regions of progressively higher relaxation time (NÉEL, 1949) and thus such experiments might be useful for providing better estimates of the importance of VRM in the geologic history of a particular sample.

Although for the Macquarie Island ophiolite we measured the VRM acquisition of only one sample from the upper pillows and one sample from the lower pillows, our data are in very good agreement with the corresponding results of BUTLER *et al.* (1976), who obtained eight VRM acquisition curves for the lower pillows and eight curves for the upper pillow.

Table 2 lists the means of the VRM components, extraporated for the entire Brunhes polarity epoch, relative to the NRM. Because the data are highly scattered (as seen from the standard errors associated with the means) the extraporated VRM was also compared to ARM (anhysteretic remanent magnetization) acquired in an 0.5 Oe direct field superimposed on and parallel to a decreasing AF of an initial value of

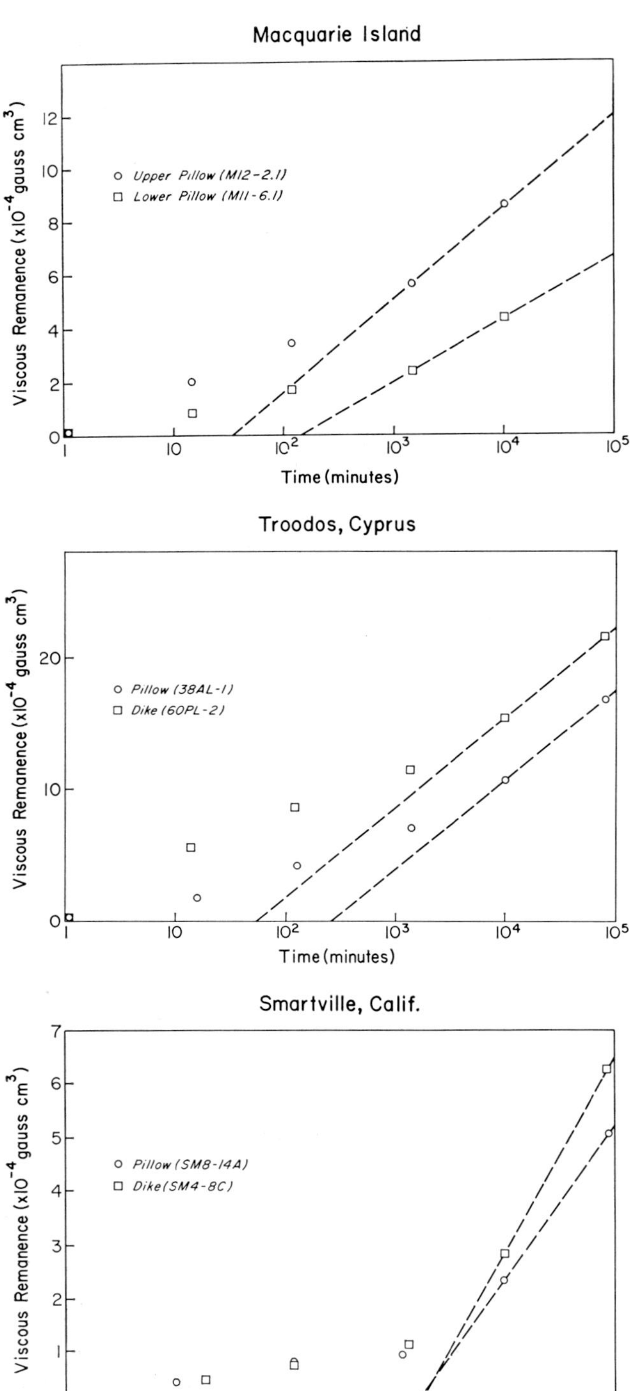

Fig. 3. VRM acquisition in an external field of 0.45 Oe for samples from the three ophiolite complexes. Since the rate of VRM acquisition usually increases with time, the slope of the last two points (dashed lines) was used to estimate the VRM contribution for the entire Brunhes epoch. The zero point for each VRM acquisition experiment is the 1,000 Oe AF demagnetization of the NRM.

Table 2.   Relative VRM contribution for the entire Brunhes polarity epoch.

Locality	Number of samples		VRM (0.7 my) / NRM $\pm E^*$	VRM (0.7 my) / ARM (0.5 Oe) $\pm E^*$
		Pillow basalts		
Macquarie Island	1	Upper pillow: low temperature oxidation	0.08	0.14
	1	Lower pillow: greenschist facies	0.55	0.21
Troodos, Cyprus	2	Zeolite facies	0.11 ± 0.01	0.074 ± 0.06
Smartville, Calif.	4	Greenschist facies	0.31 ± 0.13	0.30 ± 0.17
		Sheeted dikes		
Macquarie Island	2	Amphibolite facies	0.08 ± 0.06	0.016 ± 0.002
Troodos, Cyprus	5	Greenschist facies	0.31 ± 0.18	0.23 ± 0.08
Smartville, Calif.	7	Greenschist facies	0.21 ± 0.08	0.25 ± 0.09

* $E=\sigma\sqrt{n}$, $\sigma=$ standard error.

1,000 Oe. This was done to try to relate the VRM intensity to the specific remanence carrying ability of the sample. Despite the large scatter of both NRM and ARM-normalized data, the trend of their mean values is the same. For the upper pillows of the Macquarie Island ophiolite complex and for the pillows of the Troodos Massif the VRM contribution is seen to be least significant. However, for the metamorphosed pillows of Macquarie Island and the Smartville complex the VRM represents a substantial fraction of the samples' NRM. For the sheeted dike complexes of the Troodos Massif and Smartville, both having been altered to the greenschist facies, the VRM contributions are comparable and represent a substantial fraction of the NRM; however, the dikes of Macquarie Island, which exhibit metamorphism to the amphibolite facies, are least affected by VRM.

Thus the increased contribution of VRM to the NRM of the more highly altered pillow basalts is consistent with the results of the previous section, showing progressive degradation of the magnetic recording qualities of the pillow basalts with increased alteration. For dikes, the capability of VRM acquisition is not obviously related to their metamorphic grade.

3.4   Alternating field demagnetization

Alternating field (AF) demagnetization of NRM was done for samples from the three ophiolite complexes to assess their intensity and directional stability. Figure 4 shows representative AF demagnetization results for one pillow basalts specimen and one dike specimen from each ophiolite complex. Both the normalized intensity and the equal area stereographic projection of the NRM directions are reproduced for each specimen as function of increasing alternating fields.

The samples for the Macquarie Island ophiolite have median destructive fields (MDF) typically between 300 and 500 Oe for both pillow and dike specimens, and there is no significant difference between the MDF values of upper the lower pillows

S. Levi and S.K. Banerjee

MACQUARIE ISLAND

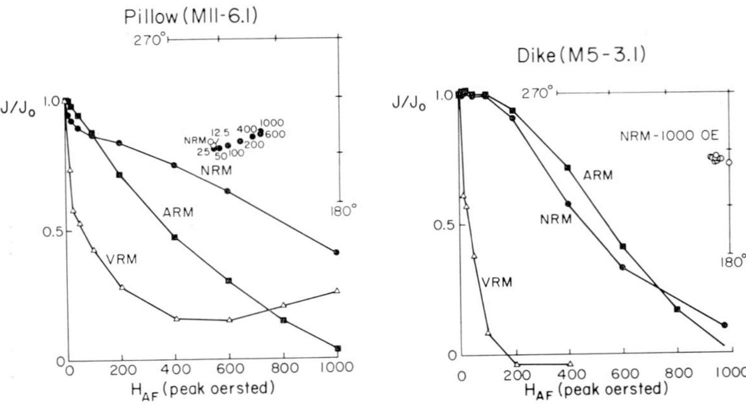

TROODOS, CYPRUS

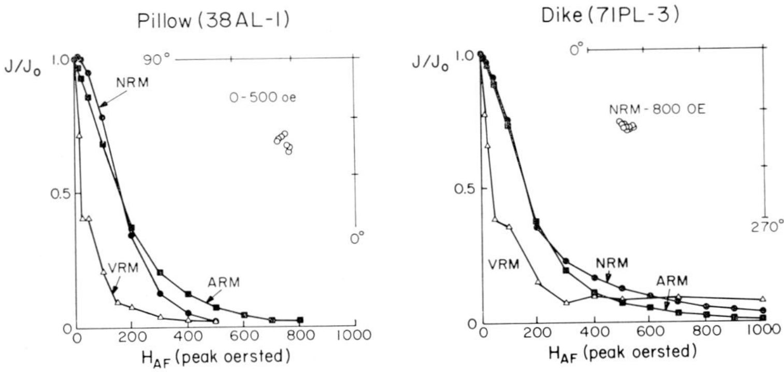

SMARTVILLE, CALIF.

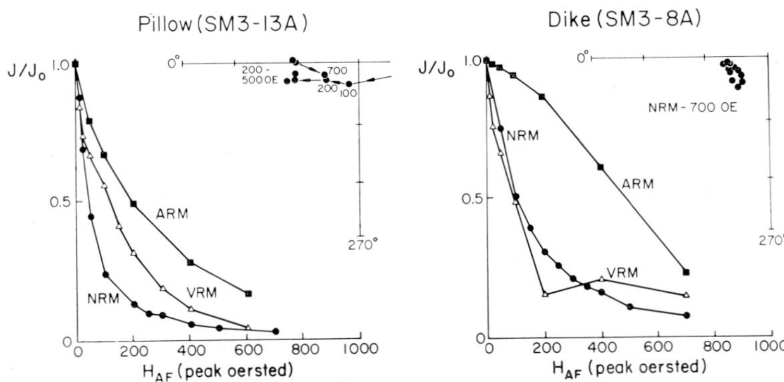

Fig. 4

(BUTLER *et al.*, 1976). The samples' stable directions are distinct from the present direction of the geocentric axial dipole at the sampling locality, and samples from a single site exhibit only small scatter in their directions. The NRM directions of samples from the upper pillow basalts and dikes are often also the stable directions but the lower pillow basalts usually contain a significant VRM component parallel to the present field (BUTLER *et al.*, 1976) which can usually be removed by AF demagnetization between 100 and 200 Oe. Because Macquarie Island is highly faulted, no meaningful between-site comparisons of the magnetization-directions could be made.

The MDFs of samples from the Troodos Massif are usually between 100 and 200 Oe for both pillow basalts and sheeted dike specimens, and the stable direction is often indistinguishable from the NRM direction. In addition, the NRM directions are distinct from the present field direction on Cyprus (VINE and MOORES, 1972). Unfortunately, all the samples which were studied by us were unoriented.

The specimens from the Smartville ophiolite, both pillow basalts and dikes, exhibit highly unstable NRM; often the MDF is not a meaningful measure of the samples' stability; whenever it is, the MDFs are usually less than 100 Oe and frequently less than 50 Oe. Although the NRM direction could sometimes be chosen as the stable direction, the stable direction is often very similar to the present direction of the geocentric axial dipole at the sampling site.

Many factors influence the stability of the remanence as measured by AF demagnetization; for example, the type of remanence, size and shape distributions of the magnetic particles, as well as their chemistry, mineralogy and oxidation states all influence the stability of the remanence. Because one or more of these parameters is usually unknown, we induced artificial remanences in samples from each site and from each lithology of the three ophiolite complexes in an attempt to determine the type of remanence represented by the NRM. The artificial remanences which were used are: VRM in a 0.45 Oe direct field applied for duration of up to 10^5 minutes; ARM produced by a 0.5 Oe direct field superimposed on and parallel to an alternating field decreasing to zero from an amplitude of 1,000 Oe; saturation IRM (isothermal remanent magnetization) produced in a direct field of 15×10^3 Oe. Typically, the ARM is more stable than the IRM, which, in turn, is more stable than the VRM.

LEVI and MERRILL (1976) have shown, for a broad range of magnetite-bearing samples that ARM and TRM (thermal remanent magnetization) produced in weak biasing fields have similar stabilities with respect to AF demagnetization. CRM (chemical remanent magnetization) can be either less stable or more stable than TRM or ARM depending on the nature of the chemical change (e.g., grain growth, oxidation or reduction), the size and shape of the particle and its chemical composition (see,

Fig. 4.   AF demagnetization of NRM, ARM and VRM for representative samples from the three ophiolites. ARM was produced by a 0.5 Oe biasing field superimposed on and parallel to a decreasing alternating field whose maximum amplitude is 1,000 Oe. VRM was produced in a 0.45 Oe field applied for between 10^4 and 10^5 minutes. Normalized intensities are plotted versus peak AF demagnetizing field. Also shown are the equal area projections of the NRM directions upon progressive AF demagnetization. Note that the appropriate quadrant of the stereogram has sometimes been rotated for snug fit.

for example, review paper by Merrill, 1975). In Fig. 4, AF demagnetization curves of NRM, ARM and VRM are shown for representative samples from the three ophiolite complexes. For the Macquarie Island ophiolite complex and for the Troodos Massif, the AF demagnetization curves of the NRM are usually much more similar to those of the ARM, the VRM demagnetization curves being distinctly less stable. This pattern is maintained for both pillow basalts and dike samples. On the other hand, for the Smartville complex , where both pillow basalts and dike samples have much 'softer' NRM, the NRM AF demagnetization curves usually more closely resemble those of the VRM than of the ARM. The strongest statement that is supported by these data is that the NRM of the Smartville pillow basalts and dikes is probably a VRM acquired during the present polarity epoch. In contrast, the NRM of the Macquarie Island samples and those from the Troodos Massif is probably not a VRM. The NRM of the zeolite facies pillow basalts of the Troodos Massif and the upper pillows of the Macquarie Island complex might represent the original TRM, subsequently modified by low temperature oxidation. The NRM of the lower pillow basalts and sheeted dikes of Macquarie Island and the greenschist pillow basalts and sheeted dikes of the Troodos Massif might be either a TRM or a CRM.

4.   Discussion

Two important assumptions are implicit whenever the magnetic properties of an ophiolite complex are used directly to construct magnetic models for the oceanic crust: 1) the ophiolite complex represents typical ocean crust originally formed at a spreading center; and 2) obduction onto land had no significant effect on the magnetic properties.

The presence of pillow basalts and overlying deep sea cherts at the three ophiolite complexes of this study are strong evidence for their marine origin. In addition, for the Troodos Massif and the Macquarie Island ophiolite the affinities of the trace elements Ti, Zr, Y to those of ocean-floor basalts seem convincing. In addition to its tectonic setting, probably the most compelling evidence for a spreading center origin of the Macquarie Island ophiolite complex is found in the Sr isopope data (Williams and Murthy, 1977). However, even for the most thoroughly examined ophiolite complex (the Troodos Massif) controversy exists as to the environment of its origin, spreading center versus island arc (e.g., Miyashiro, 1975; Moores, 1975). At present, it appears that there are no clear-cut means for resolving this controversy.

The second assumption is probably more difficult to test. However, the apparent lack of deformation of the ophiolite complexes might be taken to support the assumption of 'gentle' obduction, leading one to the conclusion that the observed greenschist or amphibolite facies metamorphism took place prior to obduction. On the other hand, DSDP basalts drilled thus far are *not* metamorphosed (beyond hydrothermal alteration and low temperature weathering) although, again, it is possible that DSDP drilling has not yet tapped deep enough levels of the oceanic crust. It might be that ophiolite complexes do not represent 'typical', relatively passive oceanic crust, but rather that they might represent regions which were tectonically more active such as areas adjacent to transform faults or adjacent to boundaries between oceanic plates and continents and thus susceptible to more extensive metamorphism.

The metamorphic grade of both the Macquarie Island complex and of the Troodos Massif increases with depth. This supports the position that metamorphism occurred on the sea floor rather than during obduction, because such alteration stratigraphy fits readily into the framework of present models of the ocean crust, which show steep temperature gradients, particularly near spreading centers (e.g., SLEEP, 1975; KUSZNIR and BOTT, 1976). However, in the Smartville complex the metamorphic grade does not appear to increase with depth, which might be used to argue that in this case additional metamorphism occurred during or after obduction.

Recognizing the serious uncertainties about the assumptions listed at the head of this section, we return for the remainder of the discussion to the framework established by these assumptions.

The pillow basalts of the three ophiolite complexes reflect progressive metamorphism increasing from the low-temperature-oxidized and the zeolitized pillows of the Macquarie Island ophiolite and the Troodos Massif to the greenschift facies pillows of Macquarie Island, Troodos Massif and the Smartville complex. In addition, recent documentation of the available data shows that the ulvöspinel content of unaltered basalts from different oceans is restricted to rather rarrow range between $0.55 < x < 0.68$, where x is the mole fraction of ulvöspinel (OZIMA et al., 1974; JOHNSON and HALL, 1977). Therefore, if it is assumed that TRM is the original NRM of the pillow basalts of these three ophiolite complexes and that their magnetic mineralogy was initially similar, then the magnetic properties of the three suites of pillow basalts might reflect their magnetic evolution with increased alteration.

The pillow basalts of the Troodos Massif and the upper pillows of Macquarie Island, which exhibit cation deficient titanomagnetites are typical of rocks dredged and drilled in the world's oceans in terms of their magnetic mineralogy as well as the characteristics of their remanence properties. In addition, rock magnetic studies of the uppermost extrusives of oceanic crust have shown that the low-temperature oxidation of the titanomagnetites occurs within only a few kilometers of the spreading centers (CARMICHAEL, 1970; IRVING et al., 1970; OZIMA et al., 1974; JOHNSON and ATWATER, 1977), and studies of DSDP rocks have shown that the cation deficient spinel phase is preserved as a metastable phase in oceanic crust as old as 100×10^6 years to depths in excess of 500 meters. It is thus likely that magnetic properties of the low-temperature-oxidized pillow basalts of the Troodos Massif and of Macquarie Island were frozen in them since shortly after they were extruded at their respective spreading centers.

The more interesting extrusives, which are present at some of the ophiolite complexes have been metamorphosed to greenschist facies and exhibit reversible thermomagnetic curves with Curie points near that of magnetite; such thermomagnetic behavior has essentially no equivalents among DSDP rocks. Whenever metamorphosed and low-temperature-oxidized extrusives are found in the same ophiolite complex (Macquarie Island and Troodos Massif), the metamorphosed pillow basalts appear to be stratigraphically lower. Heating due to burial or to contact with hot rocks or hot aqueous solution could be the cause for metamorphism. Such heating, in turn, would cause unmixing of the low-temperature-oxidized titanomagnetite to a more Fe-rich magnetic spinel, approaching magnetite, and Ti-rich, relatively non-magnetic, phases. Varne and Rubenach (1972; attributed by them to Turner, 1968), suggest that exothermic hydration reactions within the pillow basalts might be an additional heat source to effect metamorphism and consequently the unmixing of the cation-deficient titanomagnetites. The occurrence of such a thermal event is consistent with the magnetite-like Curie points of the lower pillow basalts from Macquarie Island and the greenschist facies pillows from the Troodos Massif and Smartville complex. From our present knowledge and inferences about the thermal structure of the oceanic lithosphere it seems that the most likely place for the pillow basalts to be subjected to such a thermal metamorphic event is near the spreading centers. Of course, it is possible that not all extrusive sections experience the higher temperatures required to unmix the titanomaghemites, so that in some sections of the oceanic crust irreversible thermomagnetic curves might be preserved throughout the entire extrusive column.

Although the greenshist facies metamorphosed pillow basalts are inferior magnetic recorders as compared with the zeolitized and unmetamorphosed pillow basalts, there is considerable variation between the magnetic properties of greenschist facies metamorphosed pillow basalts of the Smartville complex and the lower (greenschist facies) pillows from Macquarie Island. This is not surprising, as greenschist facies metamorphism can occur over a relatively broad pressure-temperature range. Thus differences in the magnetic properties of the metamorphosed pillow basalts could be due to differences in the pressure-temperature conditions which prevailed during metamorphism, differences in the initial bulk chemistry of the basalts, or differences in the fluids present during metamorphism. For example, the titanomagnetite grains of the lower pillow lavas of Macquarie Island are only a few microns in size, have skeletal shapes suggestive of quenched textures, and are too small for exsolution lamellae to be resolved. In contrast, the titanomagnetite grains of some of the pillows from the Smartville complex are often larger than 20 μm, their shapes are anhedral to subhedral, and no exsolution lamellae were seen. The apparent homogeneity of the grains and the apparent absence of pseudobrookite in the metamorphosed pillow basalts suggest that unmixing of the titanomaghemites must have occurred below 600°C (Carmichael and Nicholls, 1967).

The magnetic properties of the sheeted dikes cannot be readily correlated with metamorphic grade. Although the dikes have been significantly metamorphosed, their magnetic properties show relatively little variation (Tables 1 and 2). Part of the reason for this might be a result of the slower cooling history of the dikes, facilitating high temperature oxidation, so that magnetite might have been the predominant magnetic mineral when the original TRM was produced. This supposition is strongly supported by theoretical models for thermal structure at spreading centers (SLEEP, 1975; KUSZNIR and BOTT, 1976). In particular, KUSZNIR and BOTT (1976) have provided models for the isotherms in layer 3 (where the dikes occur), taking into account the release of latent heat of solidification. Their Figure 5 shows that for a spreading rate of 3 cm/y magnetite would probably be produced by high temperature oxidation and that it will acquire a pure TRM within about 330,000 years from the time of intrusion, i.e., most likely the TRM will reflect the ambient magnetic field intensity and direction of the dike intrusion epoch. However, the precise direction and intensity of the remanence will depend on the reversal frequency of the magnetic field, the spreading rate, and on the degree to which the isotherms are modified by hydrothermal circulation.

KRISTJÁNSSON (1972) obtained magnetite-like Curie points for four basalt samples recovered between 900 m and 1,932 m depth in Iceland with accompanying microscopic evidence of high temperature oxidation and KRISTJÁNSSON and WATKINS (1977) draw attention to the possible importance of such high-temperature-oxidation magnetite as a source of magnetic anomalies.

An interesting result is that the NRM intensity of the dikes from Macquarie Island is nearly two orders of magnitude lower than the NRM intensity of the other two sheeted dike complexes. In contrast to the sheeted dikes of the Smartville complex, which have abundant titanomagnetite particles, the only opaque oxide phase in the seeted dikes of Macquarie Island consists of minute (<1 μm wide) elongated magnetite grains within the plagioclase crystals, which were not affected by the metamorphism (BANERJEE *et al.*, 1974). It is possible that during metamorphism to the amphibolite facies, the original titanomagnetite grains (except those in the plagioclase) participated in the reaction(s) that produced the metamorphic amphiboles (J. Stout, personal communications, 1977).

If the marine magnetic anomalies originate predominantly from a two layer oceanic crust, where pillow basalts overlie intrusive sheeted dike complexes, then our data from three ophiolite complexes suggest that progressive alteration adversely affects the magnetic-recording qualities of the pillow basalts, rendering them progressively poorer as a source for the marine magnetic anomalies. At the same time, progressive alteration of the sheeted dikes from the same ophiolite complexes leaves their magnetic properties relatively less affected. Thus with increasing alteration of the pillow basalts the dikes become relatively more important as sources for the marine magnetic anomalies. Although from our data it is not possible to determine where in time and space the alteration actually takes place, present thermal models of the oceanic crust suggest that the most likely place for the occurrence of such depth dependent metamorphism is near the spreading center origin of the particular ocean crust.

We wish to thank Dr. J. Smewing for sending us samples from the Troodos Massif, Cyprus. J. Mellema, D. Bogdan, R. Bauer, T. Schandle and D. Prestel made many of the magnetic measurements. D. Prestel studied some petrological aspects of thin-section from the Smartville complex. We benefited from several conversations with Dr. James Stout about aspects of metamorphism of the ophiolite complexes. This research has been funded by the National Science Foundation grants DES 75-21796 and OCE 76-15255. We thank Drs. H.P. Johnson and R.T. Merrill for their thorough and helpful reviews.

REFERENCES

ATWATER, S. and J.D. MUDIE, Detailed near-bottom geophysical study of the Gorda Rise, *J. Geophys. Res.*, **78**, 8665–8686, 1973.

BANERJEE, S.K., R.F. BUTLER, and J.H. STOUT, Magnetic properties and mineralogy of exposed oceanic crust on Macquarie Island, *J. Geophys.*, **40**, 537–548, 1974.

BLAKELY, R.J., An age-dependent, two layer model for marine magnetic anomalies, *The Geophysics of the Pacific Ocean Basin and Its Margin—The Woollard Volume, Geophys. Monogr.*, **19**, 227–234, 1976.

BUTLER, R.F., S.K. BANERJEE, and J.H. STOUT, Magnetic properties of oceanic pillow basalts: evidence from Macquarie Island, *Geophys. J.R. Astron. Soc.*, **47**, 179–196, 1976.

CANN, J.R., Rb, Sr, Y, Zr and Nb in some ocean floor basaltic rocks, *Earth Planet. Sci. Lett.*, **10**, 7–11, 1970.

CARMICHAEL, C.M., The mid-Atlantic ridge near 45°N. VII. Magnetic properties and opaque mineralogy of dredged samples, *Can. J. Earth Sci.*, **7**, 239–256, 1970.

CARMICHAEL, I.S.E. and J. NICHOLLS, Iron-titanium oxides and oxygen fugacities in volcanic rocks, *J. Geophys. Res.*, **72**, 4665–4687, 1967.

DUNLOP, D.J., Theory of magnetic viscosity of lunar and terrestrial rocks, *Rev. Geophys. Space Phys.*, **11**, 855–901, 1973.

GASS, I.G. and J.S. SMEWING, Intrusion, extrusion, and metamorphism at constructive margins: evidence from the Troodos Massif, Cyprus, *Nature*, **242**, 26–29, 1973.

IRVING, E., J.K. PARK, S.E. HAGGERTY, F. AUMENTO, and B. LONCAREVIC, Magnetism and opaque mineralogy of basalts from the Mid-Atlantic Ridge at 45°N, *Nature*, **228**, 974–976, 1970.

JOHNSON, H.P., Magnetic properties of crustal rocks from the Mid-Atlantic at 23°N—DSDP Leg 45, *EOS, Trans. AGU* (abstract), **57**, 932, 1976.

JOHNSON, H.P. and T. ATWATER, A magnetic study of the basalts from the Mid-Atlantic Ridge at 37°N, *Bull. Geol. Soc. Am.*, **88**, 637–642, 1977.

JOHNSON, H.P. and J.M. HALL, A detailed rock magnetic and opaque mineralogy study of the basalts from the Nazca Plate, *Geophys. J.R. Astron. Soc.*, 1977 (in press).

JOHNSON, H.P. and R.T. MERRILL, magnetic and mineralogical changes associated with low temperature oxidation of magnetite, *J. Geophys. Res.*, **77**, 334–341, 1972.

JOHNSON, H.P. and R.T. MERRILL, Low temperature oxidation of a titanomagnetite and the implications for paleomagnetism, *J. Geophys. Res.*, **78**, 4938–4949, 1973.

KRISTJÁNSSON, L., On the thickness of the magnetic crustal layer in south western Iceland, *Earth Planet. Sci. Lett.*, **16**, 237–244, 1972.

KRISTJÁNSSON, L. and N.D. WATKINS, Magnetic studies of basalts fragments recovered by deep drilling in Iceland, and the "magnetic layer" concept, *Earth Planet. Sci. Lett.*, **34**, 365–374, 1977.

KUSZNIR, N.J. and M.H.P. BOTT, A thermal study of the formation of oceanic crust, *Geophys. J.R. Astron. Soc.*, **47**, 83–95, 1976.

LEVI, S. and R.T. MERRILL, A comparison of ARM and TRM in magnetite, *Earth Planet. Sci. Lett.*, **32**, 171–184, 1976.

LOWRIE, W., Viscous remanent magnetization in oceanic basalts, *Nature*, **243**, 27–30, 1973.

MARSHALL, M. and A. COX, Magnetism of pillow basalts and their petrology, *Geol. Soc. Am. Bull.*, **82**, 537–552, 1971a.

MARSHALL, M. and A. COX, Effect of oxidation of natural remanent magnetization of titanomagnetite in suboceanic basalt, *Nature*, **230**, 28–31, 1971b.

MERRILL, R.T., Magnetic effects, associated with chemical changes in igneous rocks, *Geophys. Surv.*, **2**, 277–311, 1975.

MIYASHIRO, A., Origin of the Troodos and other ophiolites: a reply to Moores, *Earth Planet. Sci. Lett.*, **25**, 227–235, 1975.

MOORES, E.M., Discussion of "origin of Troodos and other ophiolites: a reply to Moores," by Akiho Miyashiro, *Earth Planet. Sci. Lett.*, **25**, 223–226, 1975.

MOORES, E.M. and F.J. VINE, The Troodos Massif, Cyprus and other ophiolites as oceanic crust, evaluation and implications, *Philos. Trans. R. Soc. London, Ser. A*, **268**, 443–466, 1971.

NÉEL, L., Théorie du traînage magnétique des ferromagnétiques en grains fins avec applications aux terres cuites, *Ann. Géophys.*, **5**, 99–139, 1947.

OZIMA, M. and M. OZIMA, Characteristic thermomagnetic curves in submarine basalts, *J. Geopys. Res.*, **76**, 2051–2056, 1971.

OZIMA, M., M. JOSHIMA, and H. KINOSHITA, Magnetic properties of submarine basalts and implications on the structure of the oceanic crust, *J. Geomag. Geoelectr.*, **26**, 335–354, 1974.

PEARCE, J.A. and J.R. CANN, Ophiolite origin investigated by discriminant analysis using Ti, Zr, and Y, *Earth Planet. Sci. Lett.*, **12**, 339–349, 1971.

RYALL, P.J.C. and J.M. ADE-HALL, Laboratory-induced self reversal of thermoremanent magnetization in pillow basalts, *Nature*, **257**, 117–118, 1975.

SCHAEFFER, R.M. and E.J. SCHWARZ, The Mid-Atlantic Ridge near 45°N, IX. Thermomagnetics of dredged samples of igneous rocks, *Can. J. Earth Sci.*, **7**, 268–273, 1970.

SCHWEICKERT, R.A. and D.S. COWAN, Early Mesozoic tectonic evolution of the western Sierra Nevada, California, *Geol. Soc. Am. Bull.*, **86**, 1329–1336, 1975.

SCIENTIFIC PARTY DSDP LEG 37, Souces of magnetic anomalies of the Mid-Atlantic Ridge, *Nature*, **255**, 389–390, 1975.

SCLATER, J.G. and K.D. KLITGORD, A detailed heat flow, topographic, and magnetic survey across the Galapagos spreading center at 86°W, *J. Geophys. Res.*, **78**, 6951–6975, 1973.

SLEEP, N.H., Formation oceanic crust: some thermal constraints, *J. Geophys. Res.*, **80**, 4037–4042, 1975.

SMITH, R.E. and S.E. SMITH, Comments on the use of Ti, Zr, Y, Sr, K, P and Nb in classification of basaltic magmas, *Earth Planet. Sci. Lett.*, **32**, 114–120, 1976.

TALWANI, M., C.C. WINDICH, and M.G. LANGSETH, Reykjanes ridge crest: A detailed geophysical study, *J. Geophys. Res.*, **76**, 473–517, 1971.

VARNE, R. and M.J. RUBENACH, Geology of Macquarie Island and its relationship to oceanic crust, *Antarct. Res. Ser.*, **19**, 251–266, 1972.

VINE, F.J. and E.M. MOORES, A model for the gross structure, petrology, and magnetic properties of the oceanic crust, *Geol. Soc. Am. Mem.*, **132**, 195–205, 1972.

WILLIAMS, S. and V.R. MURTHY, Sr-isotopic and trace element geochemistry of the Macquarie Ridge ophiolite complex, *EOS, Trans. AGU* (abstract), **58**, 536, 1977.

WILLIAMSON, P., Recent studies of Macquarie Island and the Macquarie ridge complex, *Bull. Aust. Soc. Explor. Geophys.*, **5**, 19–22, 1974.

Subject Index